Damaged Ship IV

London, United Kingdom
16-17 May 2018

ISBN: 978-1-5108-8319-2

Printed from e-media with permission by:

Curran Associates, Inc.
57 Morehouse Lane
Red Hook, NY 12571

Some format issues inherent in the e-media version may also appear in this print version.

The Royal Institution of Naval Architects
8-9 Northumberland Street
London, WC2N 5DA
United Kingdom

Phone: 020 7235 4622
Fax: 020 7259 5912

publications@rina.org.uk

Additional copies of this publication are available from:

Curran Associates, Inc.
57 Morehouse Lane
Red Hook, NY 12571 USA
Phone: 845-758-0400
Fax: 845-758-2633
Email: curran@proceedings.com
Web: www.proceedings.com

CONTENTS

SOLAS 2020 PASSENGER AND CARGO SHIP DAMAGE STABILITY REGULATIONS

K W Hutchinson, Babcock International Group - Energy and Marine, UK
A L Scott, Maritime and Coastguard Agency, UK

SUMMARY

This paper commences by discussing the deliberations that have taken place at the IMO over the past decade regarding the damage stability of both dry cargo and passenger ships, including RO-PAX ferries etc. It then turns to consider and critically discuss the recent amendments to the current SOLAS2009 damage stability regulations approved by MSC 98 in June 2017, and which will enter into force for new ships constructed on or after 1st January 2020.

The degree to which ships designed to comply with the newly amended SOLAS2020 Regulations may be economically compromised in comparison with the current SOLAS2009 Regulations is yet to be fully understood. Understandably, within the industry there are currently significant concerns over the forthcoming regulatory changes particularly the substantial increase in R for passenger ships carrying up to 1,350 persons and, specifically for the operators and designers of RO-PAX ships for service in EU waters, the future of the Stockholm Agreement, to mention just two. Hence, the paper discusses the concerns of ship owners, operators and designers and how the upcoming international damage stability regulations may influence the economics of the dry cargo or passenger ship fleet.

NOMENCLATURE

A	Attained index
B	Breadth (metres)
COSS	European Committee On Safe Seas and the Prevention of Pollution from Ships
DNVGL	Det Norske Veritas - Germanisher Lloyd
EC	European Commission
EN	Explanatory Notes
EU	European Union
EMSA	European Maritime Safety Agency
Fin. MA	Finnish Maritime Administration
FSA	Formal Safety Assessment
GM_T	(Fluid) Transverse metacentric height (metres)
GOALDS	Goal-based Damage Stability EC FP-7 project
GZ	Righting lever (metres)
GZ_{max}	Maximum positive righting lever (metres) to θ_v. Not to be taken as more than TGZ_{max} in relation to the calculation of $s_{final,i}$
KG_f	Fluid vertical centre of gravity (metres)
H_S	Significant wave height (metres)
HSVA	Hamburg Ship Model Basin
IMO	International Maritime Organization
LLH	Long Lower Holds
L_S	Subdivision length (metres)
MCA	UK Maritime and Coastguard Agency
MSC	IMO Maritime Safety Committee
N	Total Number of Persons on Board, PoB
N_1	Lifeboat provision
NL MA	Netherlands Maritime Administration
NTUA	National Technical University of Athens
p_i	Probability that only one compartment(s) under consideration are flooded
PAX	Passenger
R	Required index
Range	Range of positive righting levers measured from θ_e. The positive range is taken up to θ_v (degrees) but not to be taken as more than *TRange* in relation to the calculation of $s_{final,i}$
RCO	Risk Control Options
RO-RO	Roll-On / Roll-Off
s_i	(Conditional) Probability of survival after flooding of the compartment(s) under consideration
$s_{final,i}$	(Conditional) Probability of survival in the final stage of flooding of the compartment(s) under consideration
SDC	IMO Sub-Committee on Ship Design and Construction
SDS CG	IMO SLF / SDC Subdivision and Damage Stability Correspondence Group
SLF	IMO Sub-Committee on Stability and Load Lines and on Fishing vessels (SDC from 2014)
SOLAS	Safety of Life at Sea Convention. This paper focuses only on Chapter II-1 Parts B, B-1, B-2, B-3 and B4 (subdivision and stability)
SOLAS 90	Previous version applicable to ships constructed prior to 1st January 2009 – last version based on deterministic damage stability assessment
SOLAS 2009	Current version which is applicable to new ships constructed on or after 1st January 2009 – based on probabilistic damage
SOLAS 2020	amended future version which will be applicable to new ships constructed on or after 1st January 2020 – based on probabilistic damage
SSRC	Ship Safety Research Centre
UK	United Kingdom
USA	United States of America
VMB	Vienna Model Basin
WOD	Water on Deck
θ_e	Equilibrium heel angle in any stage of flooding (degrees)
θ_{min}	Minimum heel angle in any stage of flooding (degrees) in relation to the calculation of $s_{final,i}$
θ_{max}	Maximum heel angle in any stage of flooding (degrees) in relation to the calculation of $s_{final,i}$
θ_v	Angle, in any stage of flooding, where the righting lever becomes negative, or the angle at which an opening incapable of being closed weathertight becomes submerged (degrees)

1. INTRODUCTION

Building on papers read by the authors at all three previous 'Damaged Ship' International Conferences, held at RINAs Headquarters in London in 2011 [1], 2013 [2] and 2015 [3], and also at RINAs 'Design and Operation of Ferries and RO-PAX Vessels' International Conference, held at Headquarters in 2016 [4], together with articles recently published in RINAs The Naval Architect magazine [5] and [6], this paper further considers and critically discusses the recent amendments to the SOLAS2009 Chapter II-1 damage stability regulations, for both dry cargo and passenger ships. These SOLAS2020 amendments were approved by the IMO at the 98th session of the Maritime Safety Committee, MSC 98, held at their London Headquarters between the 7th and 16th June 2017, and enter into force for new ships constructed on or after 1st January 2020.

The degree to which ships designed to comply with the newly amended SOLAS2020 Regulations may be economically compromised in comparison with those designed for compliance with the current SOLAS2009 Regulations is yet to be fully appreciated. Understandably, within the shipping industry there are at present significant concerns over the forthcoming regulatory changes, including the substantial increase in the Required Subdivision Index, R, particularly for passenger ships carrying up to 1,350 persons and also, specifically for the operators and designers of RO-PAX ships, adjustments to the survival $s_{final,i}$ factor which will tend to result in a lowering of the Attained Subdivision Index, A, making R more difficult to achieve.

There is further uncertainty for owners and operators of RO-PAX ships trading in European (EU) waters over the future of the Stockholm Agreement. A decision by the EU is awaited on whether the Stockholm Agreement will continue to be applied to new ships once the SOLAS2020 amendments come into force. A substantial EU-funded research project is currently under way to assist in the decision-making process by trying to assess to what degree the new amendments will make sufficient allowance for the issue of potential water accumulation on the vehicle deck following side collision damage involving breaching of the garage space. It is expected that the results of this research project will be available towards the end of 2018 and meetings in Brussels will then take place between the Commission and the member states to reach a decision on how to proceed.

This paper sets these regulatory changes in a historical context, showing how the standards of residual stability have steadily increased over the years, often prompted by reaction to tragic loss of life following collision or other serious incidents. Some of the underlying reasons for the changes, including the major shift from a deterministic to a probabilistic approach to damage stability assessment are outlined. The changes in the period from 2007 to the present day are then examined in some detail highlighting the differences between the SOLAS2009 amendments and those coming into force in January 2020. Finally, the paper speculates on the possible influence of the latest amendments on the design of new passenger ships constructed on or after 1st January 2020.

2. DEVELOPMENT OF SOLAS RESIDUAL DAMAGE STABILITY STANDARDS

Figure 1 illustrates how the SOLAS Convention has been gradually amended over almost six decades, since 1960, together with the associated increase the residual stability requirements for passenger ships, and dry cargo ships.

2.1 IMO SOLAS60

The SOLAS60 Convention [7] was adopted on 17th June 1960 and entered into force on 26th May 1965. It was the first major task for IMO after the Organization's creation and, bearing in mind that in those days the calculation of residual stability was a most laborious task, represented a significant step forward in safety standards given the relatively limited analysis which was economically possible before the advent of electronic computers.

At the time the intention was to keep the Convention up to date through periodic amendments, but this proved to be very slow and virtually impossible to secure the entry into force of amendments within a reasonable period of time. As a result, a completely new Convention was adopted in 1974, namely SOLAS74 [8], which included not only the amendments agreed up until that date but a new amendment procedure – the tacit acceptance procedure – to ensure that changes could be made within a specified (and shorter) period of time. The tacit acceptance procedure provides that an amendment shall enter into force on a specified date unless, before that date, objections to the amendments are received from an agreed number of parties. As a result, the 1974 Convention has been updated and amended on numerous occasions and so is still referred to today as SOLAS 1974 'as amended'.

2.2 IMO SOLAS74

Insofar as SOLAS74 residual stability requirements are concerned it will be noted from Figure 1 that there were no changes to the criteria from that given in SOLAS60.

2.3 STAB80

In the UK there was increasing demand from designers to assign numerical values to the residual GZ curve criteria and this resulted in the UK STAB80 [9] regulations which considerably increased, for example, the required GZ_{MAX} from 0.010 to 0.050 metres (in blue in Figure 1). More importantly, the numerical values allowed for the calculation of so-called 'critical' or 'limiting' KG_f / GM_T curves by computer, a much simpler methodology for a Master to determine the safety margin of a loaded ship.

2.4 IMO SOLAS90

There was a step change in the residual GZ curve criteria (in green) in the 1988 amendments which were adopted on 28th October 1988 and entered into force into force on April 29th 1990 as SOLAS90 [10 - 14]. This was largely as a result of the *MS Herald of Free Enterprise* disaster [15]. It was the loss of the *MS Estonia* [15] which acted as a spur to the introduction of the Stockholm Agreement (EC Directive 2003/25/EC, as amended) [16].

2.5 ISSUES WITH IMO SOLAS90

SOLAS90 introduced the requirement for the assessment of damaged stability to dry cargo ships for the first time, initially for ships over 100 metres L_S but later reduced to encompass ships over 80 metres. The regulations utilised the probabilistic approach utilising two draughts.

However, as with all previous amendments, SOLAS90 continued the application of the deterministic approach to the assessment of damage stability performance of passenger ships – noting that IMO Resolution A.265(VIII) [17] from 1974, which was based on the probabilistic approach and three draughts, could be applied as an equivalent assessment, but this was seldom exercised. In these deterministic regulations, damage extents were limited to two or three compartments extent longitudinally and $B/5$ penetration transversely. This obviously led to SOLAS90 'paragraph designs' being produced which were built to these limits. Hence, the result of exceeding these limits was not assessed and was therefore unknown.

In addition, the deterministic subdivision regulations included terms such as Criterion of Service, Margin Line, Floodable Length and Permissible Length etc. These were regarded as old-fashioned and rather obscure, having been in place since the first SOLAS convention following the loss of the *RMS Titanic*.

2.6 IMO SOLAS2009

It had long been realised that the determination of a passenger ship's safety level by the deterministic approach, discussed above, was rather limiting in that the assumed damage scenarios were somewhat arbitrary in terms of penetration depth, length and vertical extent. This eventually led to new set of amendments, SOLAS2009 [18, 19], based on the probabilistic approach whereby a far wider range of damage scenarios, based on a large database of actual historical damage cases is taken into account, and hence harmonising previous regulations for passenger ships and cargo ships, namely SOLAS90 [10 - 14] and Resolution A.265(VIII) [17], into one set of damage stability regulations based principally on the probabilistic approach.

In Figure 1 it will be noted that the residual GZ curve for the SOLAS2009 amendments is shown by a black dotted rather than a solid line. Whereas the GZ curve criteria for the previous deterministic regulations represent minimum values which must be attained for each assumed damage scenario, in the probabilistic approach each individual scenario does not necessarily need to achieve, for example, the GZ_{MAX} of 0.120 metres shown in Figure 1. If a given scenario does achieve all the SOLAS2009 criteria then it will be assigned a survival $s_{final,i}$ factor value of 1, implying that it will survive in all sea states up to a H_S of 4 metres. Since 99% of historical damage cases have occurred in lesser sea states this effectively means that the damage scenario in question will always survive so that it can contribute fully to A which is used to determine whether the overall safety level, as determined by R, has been achieved.

A damage scenario with, for example, a residual GZ_{MAX} of less than 0.120 metres may still be used in the summation of A but its contribution will be lessened because it's survival $s_{final,i}$ factor will be less than 1, implying in turn that the ship may not survive that particular scenario in higher sea states.

2.7 ISSUES WITH IMO SOLAS2009

Due to the equivalence of SOLAS2009 with SOLAS90, it was found that RO-PAX designs optimised to meet SOLAS2009 were found to be vulnerable to capsize following plausible damage scenarios, even in calm water. Hence, as the $s_{final,i}$ factor obviously did not include the effect of WOD, for RO-PAX ships the Stockholm Agreement [16] was retained for such ships intended for service in EU waters.

It was also found that the R index was too low, and consequently too many damage cases were permitted to result in loss of the ship in question.

During the process of developing the SOLAS2009 amendments there were many 'loose ends', that needed tidying up or clarifying, and enhancements required to the Explanatory Notes. This was undertaken post SOLAS2009 and included *inter alia* issues such as improvements to the presentation of the limiting GM_T curves, etc.

2.8 IMO SOLAS2020

As discussed above, research showed that the WOD effect was not fully accounted for in the probabilistic SOLAS2009 amendments so this is being overcome by the introduction of the so-called '20/20' amendments (shown in red in Figure 1) to the formulae for the survival $s_{final,i}$ factor [20, 3] for RO-PAX ships. This amendment is due to enter into force on January 1st 2020 and is designed to counteract the WOD effect by only assigning a survival $s_{final,i}$ factor of 1 for damages involving the garage space if the residual GZ_{MAX} is 0.200 metres and the residual range is 20 degrees in any given damage scenario.

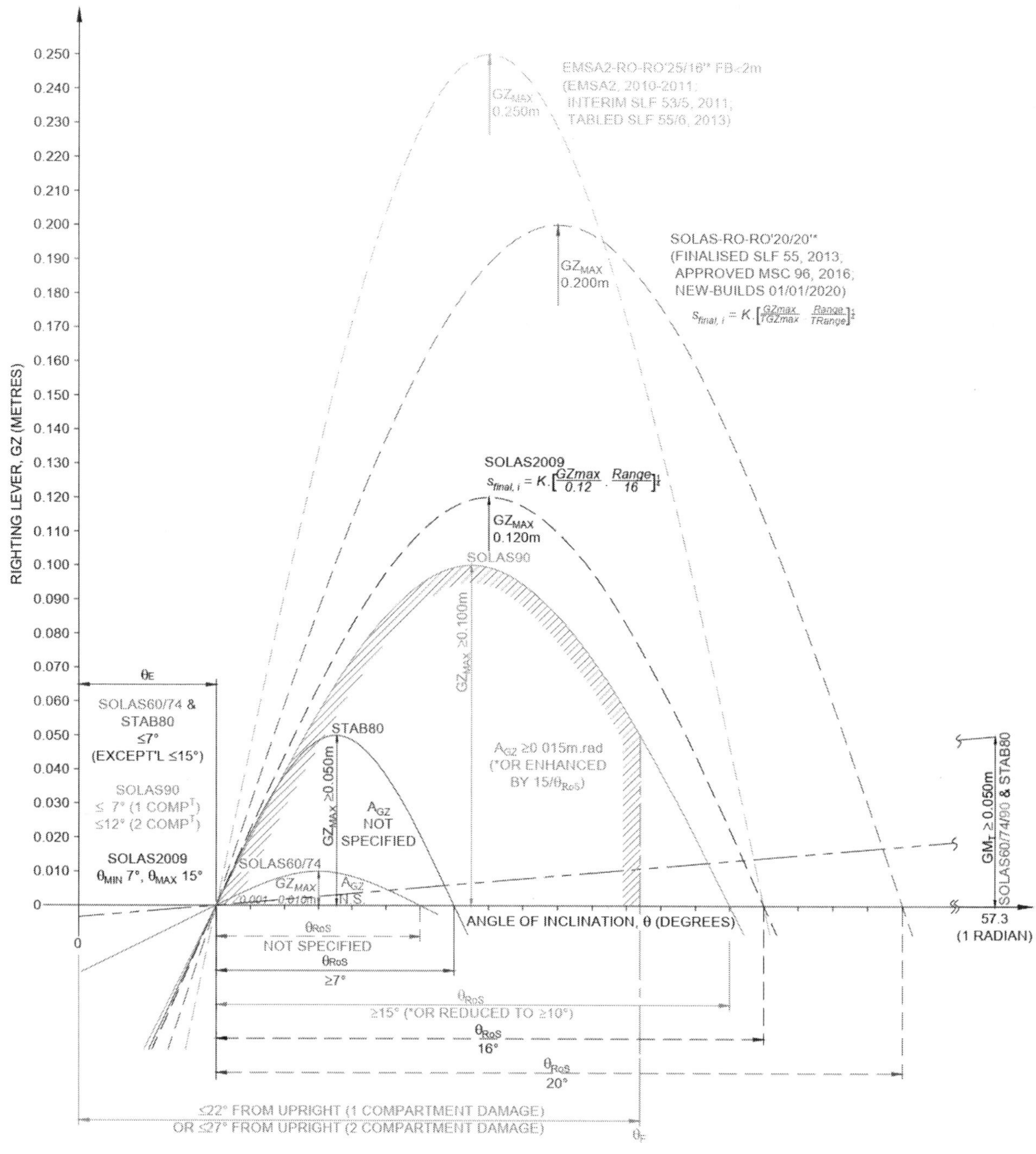

CRITERION	UNITS	DETERMINISTIC				PROBABILISTIC			
		IMO SOLAS60/74	UK STAB80	IMO SOLAS90		IMO SOLAS90 CARGO	IMO SOLAS2009 CARGO	IMO SOLAS2009 PASSENGER	IMO SOLAS2020 RO-RO PASSENGER
POSITIVE RESIDUAL RIGHTING LEVER (GZ) CURVE RANGE, θ_{RoS}	DEGREES	NOT SPECIFIED	≥7	≥15	$15 \geq \theta_{RoS} \geq 10$ (IF A_{GZ} INCREASED BY $15/\theta_{RoS}$)	20	16		20
AREA UNDER THE GZ CURVE, A_{GZ}	METRE. RADIANS	NOT SPECIFIED	NOT SPECIFIED	≥0.015	≥0.015 x $15/\theta_{RoS}$	-			
MAXIMUM RESIDUAL RIGHTING LEVER GZ_{MAX}	METRES	0.001 TO 0.010 (UK)	≥0.050	≥0.100		0.100	0.120		0.200
ANGLE OF HEEL DUE TO UNSYMMETRICAL FLOODING AFTER EQUALISATION (DAMAGE EQUILIBRIUM ANGLE), θ_E	DEGREES	≤7	≤7	≤ 7 (1 COMPARTMENT DAMAGE) ≤12 (2 COMPARTMENT DAMAGE)		MINIMUM 25 MAXIMUM 30		MINIMUM 7 MAXIMUM 15	
POSITIVE RESIDUAL METACENTRIC HEIGHT, GM_T	METRES	≥0.050	≥0.050	≥0.050		-			

Figure 1: Comparison of SOLAS Residual Stability Standards from 1960 to 2020.

3. PROCESS OF AMENDMENTS TO SOLAS

It is interesting to construct a time-line, see Figure 2 below, to illustrate how it came about that the *R* index and the $s_{final,i}$ factor used to compute the *A* index in the SOLAS2009 amendments [18, 19] had to be changed, commencing with the first research project in 2007 and culminating in the resulting SOLAS2020 amendments. Amongst other things, it shows how long it takes for amendments to be introduced and eventually to enter into force.

This is by no means a criticism of the IMO and its methods of working, which are highly effective and efficiently controlled by the secretariat. It is simply that due processes must be followed and, in the case of the SDC Sub-Committee which controls, *inter alia*, the flow of new stability regulations, the representatives of the member states who attend the annual meetings also have their main 'paid' jobs to attend to between sessions. So, unless there is a particular need to introduce new legislation quickly, in the wake, for example, of a major incident at sea, major amendments inevitably take time.

3.1 RESEARCH

Allied to this is the background research work which needs to be undertaken in order to provide evidence for the compelling need to change existing legislation. IMO does not have the funding to initiate such research, which therefore usually must come from individual member states or larger supra-national entities, such as the EU.

Such projects in themselves can take years and be technically complex in nature, often difficult to manage and very expensive. The time-line, Figure 2, attempts to illustrate the interactions over a period of years between IMO, the EC, member states, COSS and the research institutes which were required to be involved in order to change just a couple of formulae in SOLAS Chapter II-1 [18, 19], albeit formulae with wide-ranging implications for the overall safety of RO-PAX ships.

The proposed amendments were finally approved at MSC 98, in June 2017, but there is still a further waiting period before the amendments will enter into force, and hence will only be applied to new ships constructed on or after 1st January 2020. This means that, from the initial identification of the potential problem in 2007 to the enactment and entry into force of the necessary corrective measures in 2020, a period of some 13 years will have elapsed.

3.2 DISCUSSION

A concern here is that during this time period a large number of ships will have been constructed to the 'old' standard (arguably a lower one). The research projects shown on the time-line have shown almost conclusively that the original SOLAS2009 amendments provide an overall safety level equivalent to the previous, so-called SOLAS90 [10 - 14], deterministic damage stability requirements but made no provision for the so-called WOD effect which gave rise to the rapid capsize of RO-PAX ships such as the *MS Herald of Free Enterprise,* on the evening of 6th March 1987, and *MS Estonia*, in the early hours of 28th September 1994, among others [15].

Prior to January 2009, WOD was catered for in European (EU) waters by the Stockholm Agreement in association with SOLAS90 but, due to continuing doubts in the EU over the efficacy of the SOLAS2009 amendments in addressing the WOD issue, this legislation (European Directive 2003/25/EC [16]) remains in force. There is now a debate within the EU, backed up by a further large research project due to complete later in 2018, as to whether the SOLAS2020 amendments, with the increase in the *R* index and adjustments to the $s_{final,i}$ factor specifically for RO-PAX ships, will allow the Stockholm Agreement to be revoked for new RO-PAX ships constructed after 1st January 2020.

It is noteworthy that few, if any, non-European (EU) countries ever saw the need to implement either the Stockholm Agreement or their own equivalent to it for RO-PAX ships.

4. AMENDMENTS TO SOLAS

4.1 REQUIRED SUBDIVISION INDEX *R*

As discussed in Section 2.7, it was felt that the SOLAS2009 passenger ship *R* index, which basically governs the overall safety level of ships in terms of damage stability and hence the degree of subdivision to be provided, was too low. Figure 3 shows a comparison of current SOLAS2009 *R* lines (as a function of L_S, *N* and lifeboat provision) in black together with those from the GOALDS and EMSA research projects (as a function of L_S and *N* only) in gold and cyan respectively, showing evidence that the *R* index should possibly be substantially increased. There were concerns that too large an increase in *R* for smaller passenger ships, namely those with *N* up to approximately 1,500, would prove difficult for ships to attain due them being less flexibility in their subdivision arrangements, port facility constraints and a greater variety of primary design types. Equally there was a consensus that the SOLAS2009 *R* should be increased right across *N*.

4.1 (a) SDC 3, 18th to 22nd January 2016

The 'SDC 3 Proposal line' showing *R* versus *N* agreed at SDC 3 and sent to MSC 96 for approval is shown in pink in Figure 3. By way of example, for a small *N* of 1,000 RO-PAX the *R* would increase from the SOLAS2009 values of between 0.720 and 0.745 (depending on the lifeboat capacity) to 0.837 under this proposal, and for a ship with a *N* of 2,000 the *R* would increase from between 0.755 and 0.791 to 0.861.

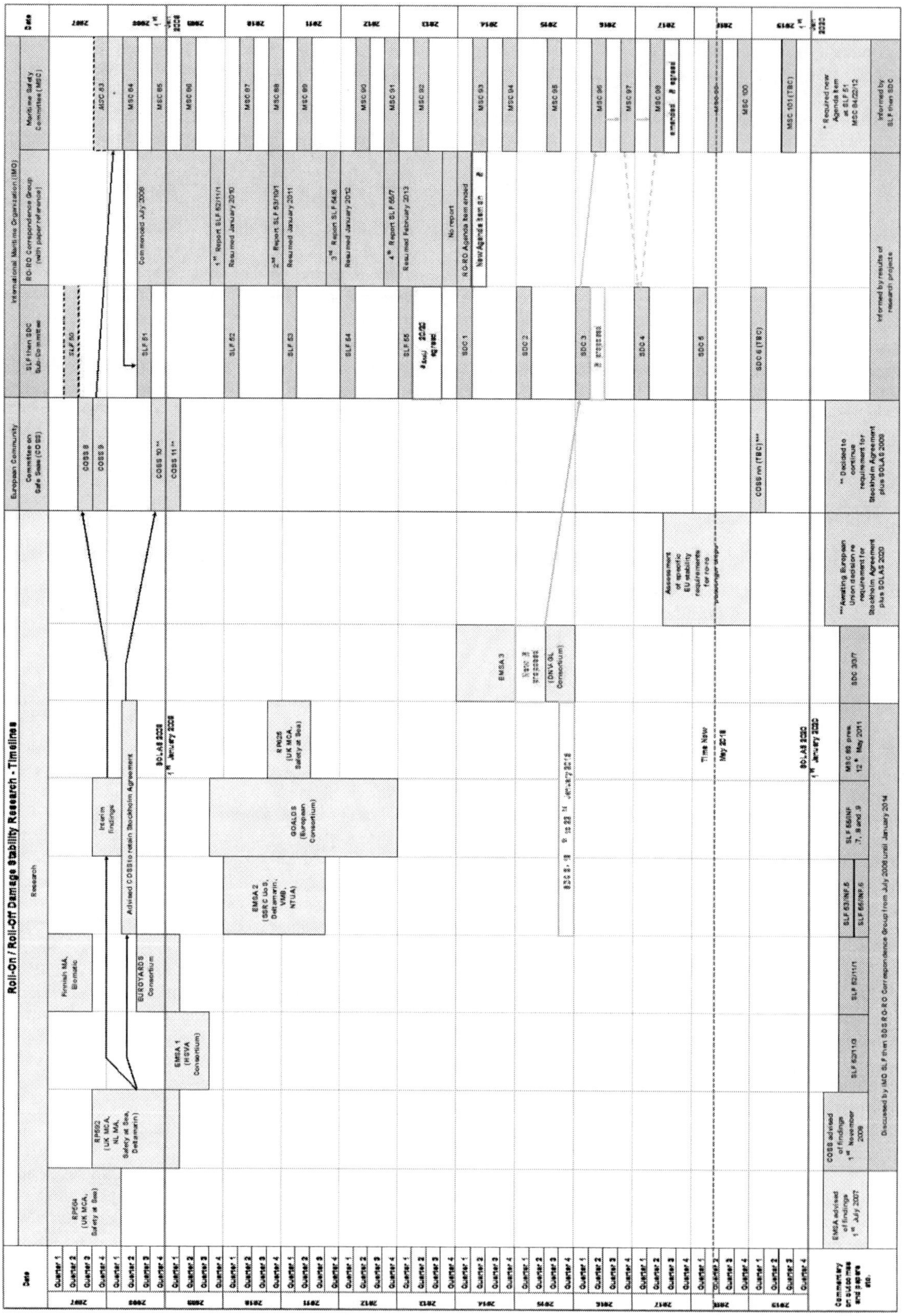

Figure 2: Time-line showing development of SOLAS Amendments regarding Passenger Ship Damage Stability.

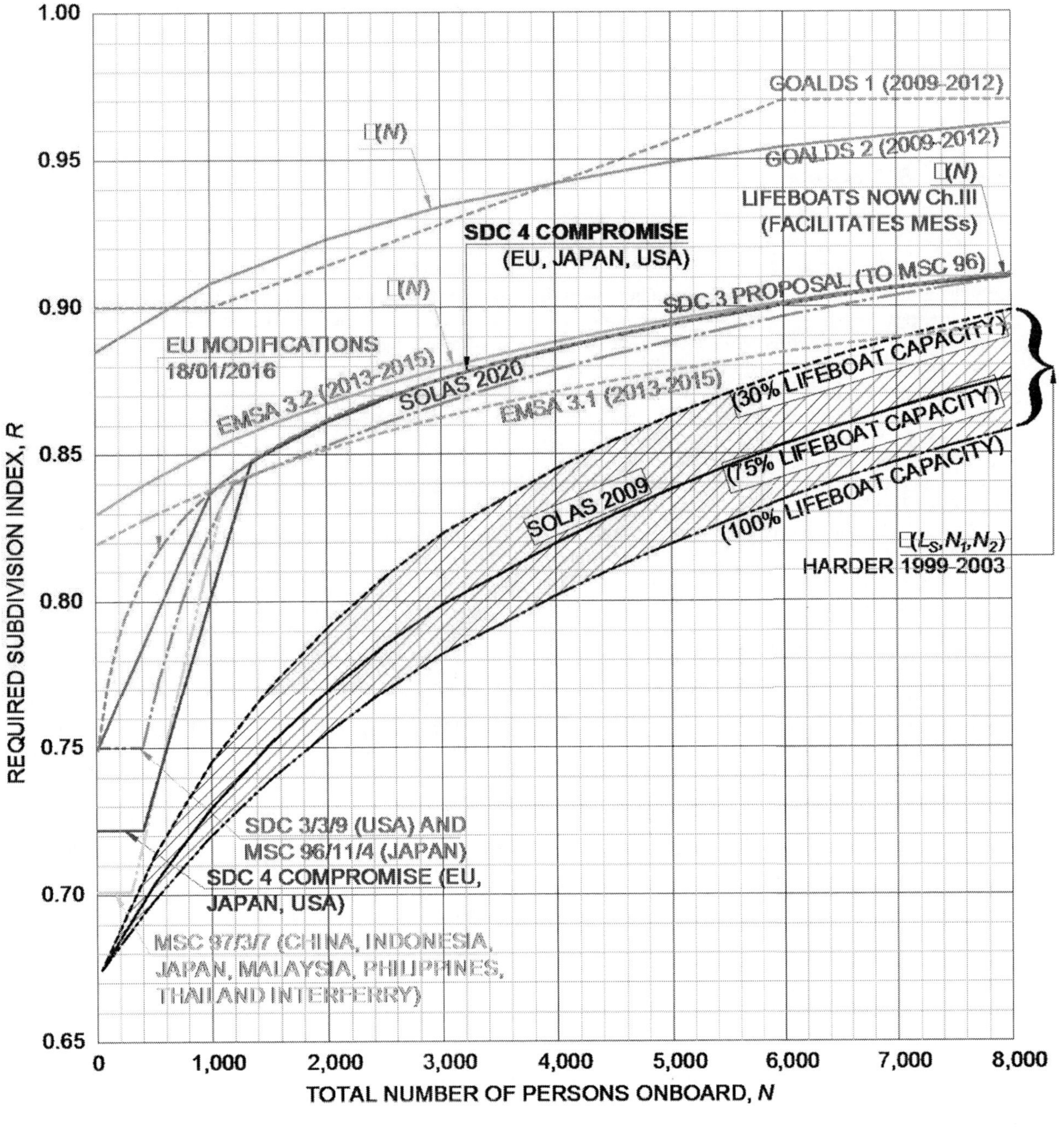

Figure 3: Comparison of current, proposed and amended Required Subdivison Index *R*.

4.1 (b) MSC 96, 11th to 20th May 2016

At MSC 96 there were significant objections, particularly for smaller passenger ships - typically defined as those with less than approximately N of 1,000. Japan proposed the green chain-dotted 'MSC 96/11/4 line', for which the R is 0.824 for a N of 1,000 and R is 0.853 for a N of 2,000. The EC and EU member states had anticipated these objections and hence prepared counter-arguments - note the 'EU 18/01/2016' line also shown in green but dashed. MSC 96 noted the above arguments but nonetheless approved the complete set of amendments, including the 'SDC 3 Proposal line' for adoption at the following MSC 97

4.1 (c) MSC 97, 21st and 25th November 2016

At MSC 97 a further significant paper was submitted by China, Indonesia, Japan, Malaysia, the Philippines, Thailand and Interferry proposing to modify the 'SDC 3 Proposal line' agreed at SDC3. This yellow 'MSC 97/3/7 line' suggested that the R should be 0.808 for a N of 1,000 and a R of 0.861 for a N of 2,000 - the latter being unchanged from the 'SDC 3 Proposal line', therefore illustrating that their primary concern was to reduce the R index for ships with fewer persons on board. After extensive debate, it was decided to hold adoption of the draft amendments in abeyance until MSC 98 in order to facilitate further discussions etc. – noting that this would still allow the full raft of SOLAS2020 amendments to enter into force on 1st January 2020.

4.1 (d) SDC 4, 13th to 17th February 2017

The debate then shifted to SDC 4 where a further compromise was reached namely the 'SDC 4 Compromise line' which is shown in blue in Figure 3, which proposed a R of 0.816 for a N of 1,000 and maintaining a R of 0.861 for a N of 2,000.

4.1 (e) MSC 98, 7th to 16th June 2017

The red 'SOLAS 2020' line was finally agreed, namely: maintaining R constant at 0.722 for a N up to 400 (this being the Chapter II-1 Regulation 8 breakpoint between 1- and 2-compartment deterministic minor damages); then increasing it linearly to intersect the line previously agreed at MSC 96 (largely as the result of the EMSA 3 research project) at an R of 0.8473 and a N of 1,350 (this latter figure was itself a compromise between setting a breakpoint at a N of 1,100 or a N of 1,600 to determine the slope of the line connecting it to a N 400); and thereafter following MSC 96 line (based on EMSA 3).

4.1 (f) Influence of Lifeboat provision

From the formulae given in Figure 4, and as indicated in Figure 3 for the proposed R lines, the current link in SOLAS2009 between lifeboat provision (N_1) and R has been removed in SOLAS2020.

It was felt that some passenger ships with a relatively short length, L_S, could carry a disproportionately large number of persons, and vice versa, so that it was preferable to simplify the formula by making R solely dependent on N.

In addition, there are now several equally effective means of rescue available other than lifeboats. Therefore, it was decided to leave the issue of minimum lifeboat provision solely to the Chapter III SOLAS Regulations on life-saving appliances and arrangements rather than linking it to subdivision and damage stability.

Whilst it is undoubtedly the case that lifeboats provide a degree of re-assurance to passengers, the actual proportion of passengers rescued after a damage incident will depend on circumstances other than just the number of lifeboats available, amongst which are, for example: the prevalent weather conditions; whether it is day or night-time; and, the availability of local search and rescue facilities.

4.1 (g) Comparison of SOLAS2009 and SOLAS2020

For reference, R is given in SOLAS Chapter II-1 Regulation 6.

The specific formulae for dry cargo ships, which is a function of L_S only, is given in Regulations 6.2.1 and 6.2.2. As can be seen from the green lines and text given in Figure 4, that there are no changes for dry cargo ships proposed for SOLAS2020 over that existing for SOLAS2009.

The specific formulae for passenger ships appear in Regulation 6.2.3, and from Figure 4 it can be seen that R is a function of both L_S and N – noting that the values for L_S on the lower x-axis of Figure 4 are purely typical ones from a regressed mean relationship against N, on the upper x-axis, for the world fleet of passenger ferries and cruise ships. The net result in terms of a percentage increase in R for the red 'SOLAS2020 Passenger' line in comparison with the black 'SOLAS2009 Passenger' line assuming 75% lifeboat capacity, N_1, it can be seen that the most significant increase lies in the range between N of 1,350 and 2,000 which roughly covers the medium to larger-size RO-PAX ships. It must be remembered that R is also a measure of how many of the whole array of damage cases and sea states encompassed by the probabilistic method are permitted to capsize. Thus, very approximately, if R is 0.75 then 1 in 4 of all cases may result in capsize whereas for an R of 0.80 the expected loss rate is reduced to 1 in 5 cases. An increase in R from 0.75 to 0.85 therefore represents a reduction in the total number of damage cases predicted to result in loss from approximately 1 in 4 to1 in 6.

From this it can be concluded from this that a very significant increase in safety level will have been put in place for all new passenger ships from 1st January 2020.

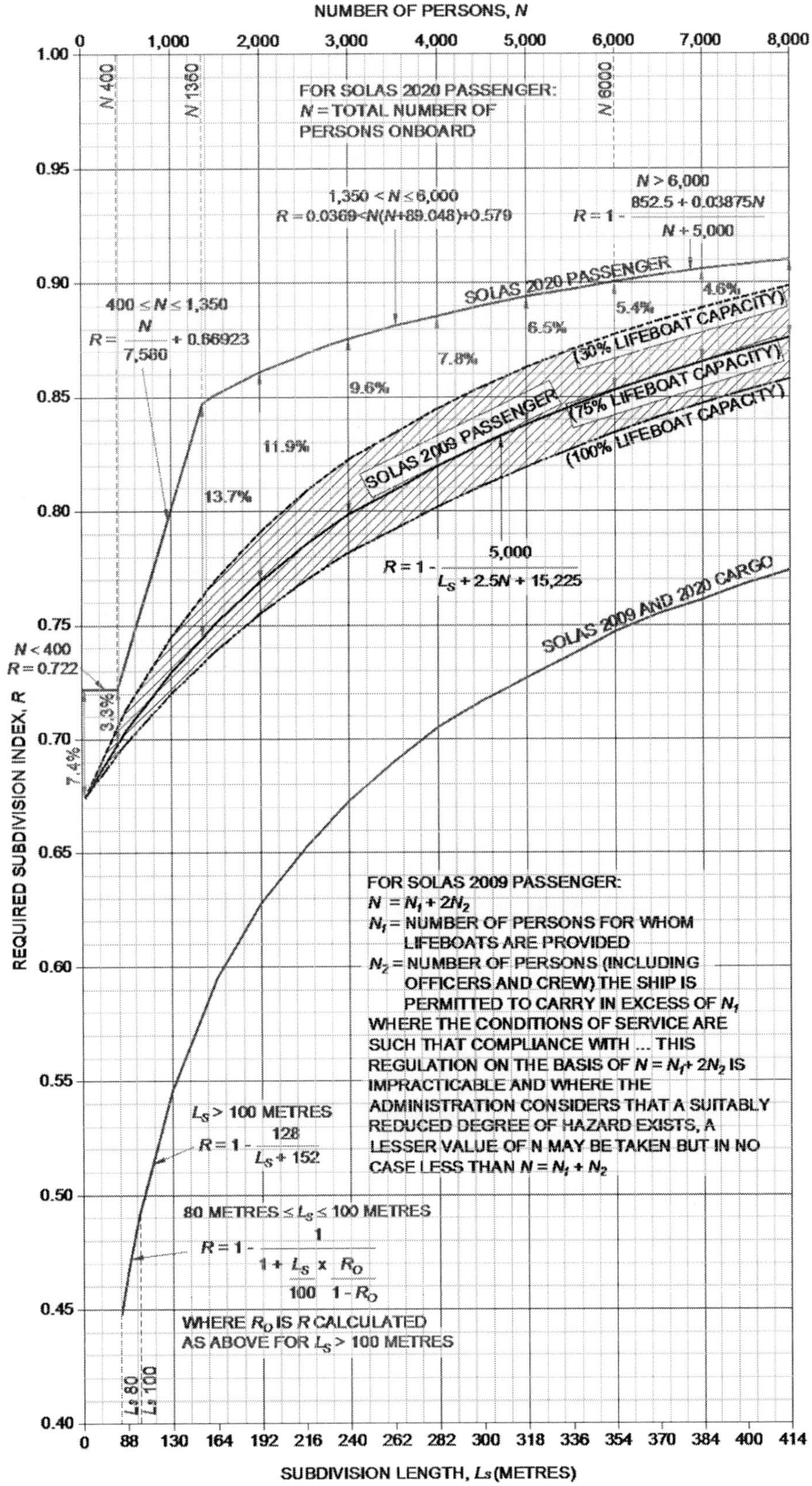

Figure 4: Comparison between the Required Indices R for SOLAS 2009 and 2020.

4.2 PROBABILITY OF SURVIVAL IN FINAL STAGE OF FLOODING $s_{final,i}$ FACTOR

For RO-RO passenger ships, in addition to the above modifications to *R*, there is also a change in the so-called $s_{final,i}$ factor in SOLAS Chapter II-1 Regulation 7-2.3. This means that the *A* index will be lowered to some degree resulting in the *R* index becoming more difficult to achieve.

This change is aimed at accounting for the increased vulnerability of RO-RO passenger ships to the WOD effect whereby flooding of the open vehicle deck can bring about rapid capsize, something which as discussed previously in Section 2.7 was not considered in the SOLAS2009 amendments. SOLAS2009 Chapter II-1 Regulation 7-2.3 has now been replaced by the following in SOLAS2020:

$$s_{final,i} = K \cdot \left[\frac{GZ_{\max}}{TGZ_{\max}} \cdot \frac{Range}{TRange} \right]^{\frac{1}{4}}$$

where:

GZ_{max} is not to be taken as more than TGZ_{max};
Range is not to be taken as more than *TRange*;
TGZ_{max} = 0.20 m, for RO-RO passenger ships each damage case that involves a RO-RO space,
= 0.12 m, otherwise;
TRange = 20°, for RO-RO passenger ships each damage case that involves a RO-RO space,
= 16°, otherwise;
$K = 1$ if $\theta_e \leq \theta_{min}$
$K = 0$ if $\theta_e \geq \theta_{max}$

$$K = \sqrt{\frac{\theta_{\max} - \theta_e}{\theta_{\max} - \theta_{\min}}} \quad \text{otherwise,}$$

where:

θ_{min} is 7° for passenger ships and 25° for cargo ships; and
θ_{max} is 15° for passenger ships and 30° for cargo ships.

As illustrated in Figure 5, the changes in the GZ_{max} and *Range* requirements for damage cases involving the RO-RO space represent a substantial increase in comparison with SOLAS2009 implying that the $s_{final,i}$ factor for an identical damage / loading scenario could be significantly reduced when computed using the new SOLAS2020 formula. This in turn could mean that the partial Attained Indices A_s, A_p and A_l in Regulation 7 will be reduced as will the overall *A*, resulting in *R* being more difficult to attain. One mitigating factor is that the reduction in the partial indices is likely to become less as draught decreases and, for the light draught, d_l, A_l may not be affected at all thanks largely to the overall average increase in residual freeboard.

Using the SOLAS2020 formula the likelihood is that those cases which comfortably achieved an $s_{final,i}$ factor of 1 under SOLAS2009 will continue to do so and those vulnerable to capsize with an $s_{final,i}$ factor of 0 will continue to be so. In the intermediate range of $0 \leq s_{final,i} \leq 1$, the values will decrease for all cases involving damage to the vehicle space. Some will be reduced sufficiently to indicate that capsizing may now occur ($s_{final,i} = 0$) and those with a marginal $s_{final,i}$ factor of 1 could be reduced to something less than 1, indicating an increased risk of capsize in heavier seas.

4.3 MINOR DAMAGE

As only a few text amendments have been made regarding minor damage for SOLAS2020, the general idea that a passenger ship should be able to withstand a limited extent of side damage anywhere along its length will remain in force. Hence, the minor damage regulations (Regulation 8 of SOLAS 2009 Chapter II-1) will continue to apply to SOLAS2020 ships.

4.4 EXPLANATORY NOTES

Apart from the changes to SOLAS Chapter II-1 Part B-1 Regulations 6 and 7-2.3 discussed above, Figures 6 and 7 summarise the most significant remaining regulatory amendments and changes to the Explanatory Notes which will come into force on 1st January 2020.

5. LIKELY EFFECT OF SOLAS2020

5.1 PASSENGER SHIPS

For passenger ships, typically cruise liners, it is not expected that many difficulties will be experienced in meeting the new *R* index as the increase, especially for the larger ships, is relatively small.

5.2 RO-PAX SHIPS

To the autors' knowledge, no new RO-PAX designs optimized to meet the proposed changes to the *R* index and $s_{final,i}$ factor have yet been produced. Hence, at this stage comments can only be speculative in nature. The changes to the $s_{final,i}$ factor will tend to reduce the *A* index as fewer damage cases will achieve a positive $s_{final,i}$ contribution. Some studies have shown that the *A* index will only be reduced by approximately 1% to 3% for medium-sized to larger RO-PAX ships but a recent study by Denmark (see SDC3/INF.4 [21, 22]) shows that the reduction in *A* can be as much as 10% to 12% for smaller ships which, if confirmed, is quite significant.

For the *R* index, Figure 3 shows that for a *N* of 2,000 RO-PAX, the *R* index will increase from around 0.78 to 0.86 which is significant but, according to the EMSA3 [23] study, this is achievable whilst still allowing the ship to operate economically.

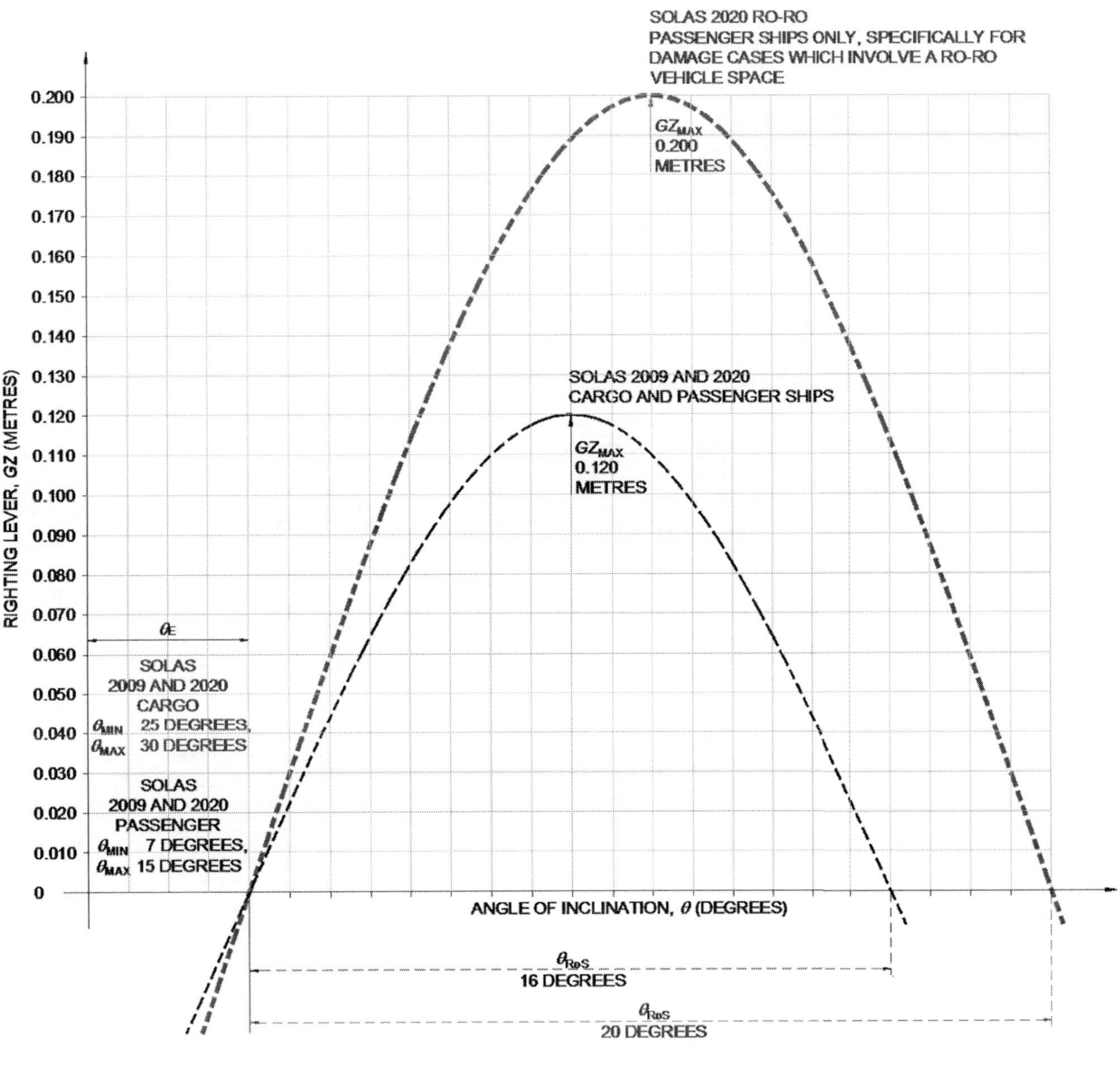

CRITERION	UNITS	SOLAS 2009 AND 2020 CARGO	SOLAS 2009 AND 2020 PASSENGER	SOLAS 2020 PASSENGER RO-RO SPACES
POSITIVE RESIDUAL RIGHTING LEVER (GZ) CURVE RANGE, θ_{RoS}	DEGREES	16		20
MAXIMUM RESIDUAL RIGHTING LEVER GZ_{MAX}	METRES	0.120		0.200
ANGLE OF HEEL DUE TO UNSYMMETRICAL FLOODING AFTER EQUALISATION (DAMAGE EQUILIBRIUM ANGLE), θ_E	DEGREES	MINIMUM 25 MAXIMUM 30	MINIMUM 7 MAXIMUM 15	

Figure 5: Comparison between SOLAS 2009 and 2020 *GZ* curve requirements to give an $s_{final,i}$ of 1.

Regulation	Subject	Amendment	Reason
A 1.3.4	Cargo ship alterations	Deleted	*A/R* ratio not relevant for new SOLAS 2009 ships
A 2	Definitions	Draught, trim, bulkhead deck modified	Increased clarity
B 4	General applicability	Footnote excluding specified cargo ships now in Regulation text	Footnotes do not have same legal authority as Regulation text
B-1 5-1	Information to the Master	Limiting *KG / GM* data now to include trim	Expansion and clarification, particularly of Reg. B-1 5-1/4
B-1 7.2	Attained Index *A*	Treatment of trim in calculation of *A*	Increased clarity
B-1 7.3	Residual *GZ* calculation	Constant displacement removed; free trimming added	Added weight method can be used during IS flooding
B-1 7-2.5	Bulkhead deck	"of passenger ships and the freeboard deck of cargo ships"	Clarification - applies in many Regulations
B-1 7-2.5.3.3	Immersion giving $s = 0$	Immersed pipe / duct in damage extent if risk of progressive flooding	Expansion and clarification
B-1 7-2.5.5	Watertight hatch covers	Reference to "small watertight hatch covers" removed	"Small" undefined; "watertight" now applied to all hatch covers
B-1 8.1 and 2	Fore end damage (passenger ships)	Text changes and how to account for trim	Clarification of extent of fore end damage
B-1 8.3	Extent of damage (passenger ships)	Removes references to *N* and L_S; now "persons carried"	Consequential upon changes to *R* formula in Reg. B-1 6.2.3
B-2 9.3 and 6-8	Double bottoms	Small wells, vertical damage extent; cargo ships < 80 metres	Expansion and clarification
B-2 12.5.1	Fore Peak bulkhead	Penetration pipe may be fitted with a butterfly valve; cargo ships	Consolidates a common practice into SOLAS
B-2 16	Testing of watertight closures	Title changed; hatches to be tested; includes small cargo ships	Testing of hatches omitted previously
B-2 16-1.2	Vents through the bulkhead deck	Added "watertight area"; removed "allowable" and "IS"	To harmonise with the rest of probabilistic SOLAS
B-2 17.3	Air pipes within a superstructure	Now considered unprotected if no watertight means of closure fitted	Consequential upon switch to probabilistic SOLAS
B-4 19.2	Open watertight doors in passenger ships	Deleted	Provision now conditional; see MSC.1/Circ.1380
B-4 19-1	Damage control drills	Whole new provisions for passenger ships	Consequences of *Costa Concordia* loss
B-4 20	Loading of passenger ships	"Passenger" deleted	Now applies to all ships
B-4 20.1	Loading of ships	Ship to be upright; loading in accordance with SIB permitted	Initial list not permitted; confirmation of current practice
B-4 21.1 and 4	Operation of watertight doors	"drills" replaced with "operational tests"	More appropriate wording
B-4 21.1	Leaving port	"leaving port" replaced with "voyage commences"	"Port" is too specific; replaced in several Regulations
B-4 22.3	Open watertight doors	Footnote added referring to new guidance in MSC.1/Circ.	See B-4 19.2. Updated Circ.1380 not yet available
C 35-1 3.4	Bilge pumps	"All flooding conditions" now limited by reference to Reg. B-1 8	Clarification

Figure 6: Main amendments to SOLAS 2009 Chapter II-1 Parts A, B, B-1, B-2, B-4 and C (MSC.421(98)).

Explanatory Note	Subject	Amendment	Reason
A 1.3	Application	Much expanded; explains what happens following alteration	Clarifies SOLAS text now that Reg. A 1.3.4 is deleted
A 2.11	Light service draught	Reference to lower draught limit deleted; excludes temporary water ballast	Moved to EN B-1 5-1.4; excludes BWM, dry-docking etc.
B 4.1	General applicablity	Entire table and footnote on OBO ships deleted	All footnotes now included in SOLAS text
B-1 5.2	Intact Stability	Reference to MSC/Circ.1158 deleted; some contents moved here	Updated guidance on what constitutes a sister ship
B-1 5.4 and 5	Alterations	New guidance included	Clarification of when to incline and when new SIB needed
B-1 5-1.4-6	Information to the Master	Completely revised and expanded, with new diagrams	Clarification; dealing with trim in *GM / KG* limit curves etc.
B-1 6.2.4	Degree of hazard	Deleted	*R* now dependent only upon *N*, persons on board
B-1 7.1	Attained Index *A*	New EN's 7 to 10 added	Clarification on when new *A* calculations are needed
B-1 7.2	Effect of trim on *A*	Much reduced here but moved to EN B-1 5-1.4-6 in a modified form	More rational to included these notes under EN B-1 5-1
B-1 7.5	*A* calculation at ends	New EN3 inserted, with diagram; old EN3 now EN4	Guidance on treating penetration with curved waterlines
B-1 7.6	Free surfaces	New EN included with cross-reference to EN for *R* under Reg. 7-2.2	How to deal with free surfaces in all stages of flooding
B-1 7.7.1	Pipes / valves near bulkhead deck	Now relaxes "directly adjacent" to "as close as practicable"	Clarification; occurs in several places
B-1 7.7.2	Minor progressive flooding	Changes pipe cross-sectional area; new breakpoint 150 metres	Different requirement needed for larger ships
B-1 7-1.1.2	Transverse subdivision	Two new ENs with diagrams added (4.1 and 4.2)	Clarification of *b*; treatment of complex waterlines aft
B-1 7-2.2.3	Sequential flooding	Extensively expanded with new sub-paragaphs and examples	Clarification of complex calculation procedures needed
B-1 7-2.2.4	Cross-flooding	Extensively expanded; includes alternative procedures	New MSC.362(92) replaces MSC.245(83); clarification
B-1 7-2.3	RO-RO $s_{final,i}$	New EN inserted	Explains use of T_{GZmax} and T_{range}
B-1 7-2.4	Displacement to use	Deleted	Now in revised SOLAS text
B-1 7-2.4.1.1	Definition of breadth	Deleted	Now in revised SOLAS text
B-1 7-2.5	Cross-flooding in cargo ships	Deleted	Now in revised SOLAS text for Reg. B-1 7-2.2
B-1 8.3.2/5	Persons on board	Deleted	*R* now dependent only upon *N*, persons on board
B-2 9	Double bottoms	Extensively expanded and modified with new diagrams etc.	Reflects changes in the SOLAS text and expands it
B-2 12	Bow doors	Deleted	Now incorporated into SOLAS text for Reg. B-2 12.7 and 12.8
B-2 12.6.1	Butterfly valves	New EN inserted with diagrams	Reflects changes in the SOLAS text
B-2 12.10/11	Aft Peak bulkheads	New ENs inserted with diagrams	Aft end bulkhead and stern tube arrangements
B-2 13.2.3	Closed / open pipes	New EN inserted	Clarification of what constitutes open / closed systems
B-2 13.7.6	IEC standard	Deleted	Footnote corrected in SOLAS text
B-2 15-1	Cargo ship openings	New EN inserted	Clarifies applicability of Regs. B-2 15-1.1 to 15-1.3
B-2 17.1	Sliding watertight doors	Extensively expanded and modified with new diagrams etc.	Necessary clarification of terminology
B-2 17.3	Air pipe open ends	Revised	Reflects changes on SOLAS text
B-2 17-1	RO-RO access	New EN inserted	Clarifies applicability of Regs. B-2 17-1.1.1 to 17-1.1.3
B-4 22	Use of "Port"	New EN inserted	Explains wider use of the term "port"

Figure 7: Main amendments to the Explanatory Notes (SDC4/16 Annex 1, MSC.429(98)).

What is clear at present is that new generation RO-PAX ships, constructed after the SOLAS2020 amendments come into force on 1st January 2020, will be much safer but less economically competitive than the current designs. This leaves a risk that the life-span of existing tonnage will be extended as far as possible into the future as new-building orders are delayed. Much will depend on the skill of designers in minimising the economic dis-benefits of the increased safety level by careful selection of the RCO's such as increased freeboard or *B*.

The overall economic climate will also be important. If rates are good and prospects promising, then new ships will inevitably be ordered anyway. At least there will be a 'level playing field' for all ships constructed after 2020 but, as previously, there is no intention of applying the amendments retrospectively to existing ships so there will always be an imbalance between the pre and post 2020 RO-PAX ships until the older ships are slowly withdrawn from service.

Anathema to RO-RO owners and operators is the RCO of utilising watertight doors on the main car deck to achieve a high attained *A* index. It is not believed that this will be necessary, whereas the much higher levels of the *R* index proposed in earlier research work may have left this as the only option.

5.2 LONG LOWER HOLD RO-PAX SHIPS

A concern to the owners and operators of RO-PAX ships with LLHs is whether it will still be possible to design such ships using the SOLAS2020 amendments. It is to be expected that the higher *R* index will result in the shortening of the LLH but whether this will render the design economically unviable is not clear present.

5.4 EUROPEAN RO-PAX SHIPS

If the Stockholm Agreement [16] is dispensed with for new RO-PAX ships constructed after 1st January 2020 then, in theory, the provision of $B/5$ longitudinal bulkheads to afford a degree of protection to the LLH from side damage will no longer be necessary. A RO-PAX ship fitted with a LLH could then, for example, be optimized to meet SOLAS2020 by fitting $B/10$ longitudinal bulkheads, making the LLH increasingly vulnerable to flooding following side damage. It is expected that this issue will be one of the main considerations of the current EU research project and could be one of the main deciding factors as to whether the Stockholm Agreement will be retained after 1st January 2020.

6. CONCLUSION

This paper has discussed the regulatory amendments agreed at IMO regarding damage stability for SOLAS ships constructed on or after 1st January 2020.

It is difficult at present to predict whether the undoubted improvements to safety the SOLAS2020 changes will provide will be sufficient to allow the European regulatory authorities to revoke the (deterministic) Stockholm Agreement for new RO-PAX ships. There is an acknowledged need for research into whether the increases in R and $s_{final,i}$ will allow the Stockholm Agreement to be revoked and hence total reliance placed on the probabilistic SOLAS2020 regulations.

7. DISCLAIMER

The views expressed in this paper are those of the authors and do not necessarily represent those of the organisations with which they are affiliated and the professional institutions of which they are members.

8. REFERENCES

1. HUTCHINSON, K.W., SCOTT, A.L., WRIGHT, P.N.H., WOODWARD, M.D. and DOWNES, J., 'Consideration of Damage to Ships from Conceptual Design to Operation: The Implications of Recent and Potential Future Regulations regarding Application, Impact and Education', *Proceedings of the International Conference on The Damaged Ship*, The Royal Institution of Naval Architects, London, United Kingdom, 26th and 27th January, 2011. ISBN: 978-1-905040-79-7

2. HUTCHINSON, K.W. and SCOTT, A.L., 'The Development and Impact of Current and possible Future International Damage Stability Regulations', *Proceedings of the International Conference on The Damaged Ship II*, The Royal Institution of Naval Architects, London, United Kingdom, 30th and 31st January, 2013. ISBN: 978-1-909024-11-3

3. HUTCHINSON, K.W. and SCOTT, A.L., 'Current and possible future Intact and Damage Stability Passenger Ship Regulations, specifically the provision of Damage Stability Information and Verification Tools for the Master', *Proceedings of the International Conference on The Damaged Ship III*, The Royal Institution of Naval Architects, London, United Kingdom, 25th and 26th March, 2015. ISBN: 978-1-909024-38-0

4. HUTCHINSON, K.W. and SCOTT, A.L., 'Passenger RO-RO Ferry Damage Stability: Status and Development of International Regulations', *Proceedings of the International Conference on The Design and Operation of Ferries & Ro-Pax Vessels*, The Royal Institution of Naval Architects, London, United Kingdom, 25th and 26th May, 2016. ISBN: 978-1-909024-54-0

5. HUTCHINSON, K.W. 'Damage stability update', The Naval Architect, The Royal Institution of Naval Architects, London, United Kingdom, May 2017. ISSN: 0306-0209

6. HUTCHINSON, K.W. and SCOTT, A.L., 'Damage stability: upcoming SOLAS 2020 regulations', The Naval Architect, The Royal Institution of Naval Architects, London, United Kingdom, September 2017. ISSN: 0306-0209

7. *International Convention for Safety of Life at Sea, 1960*, Treaty Series No. 65 (1965), Her Majesty's Stationary Office, London, United Kingdom, 17th June, 1965.

8. *International Convention for Safety of Life at Sea, Consolidated text of The SOLAS 1974 Convention, The 1978 Protocol, The 1981 and 1983 SOLAS Amendments*, International Maritime Organization, London, United Kingdom, 1986. ISBN 92-801-1200-7

9. MSN 1698 (M) *The Merchant Shipping (Passenger Ship Construction: Ships of Classes I, II and II(A) Regulations 1998,* Maritime and Coastguard Agency, Southampton, United Kingdom, December 1998. (MS 26/7/33)

10. SOLAS Consolidated Edition, 1992, *Consolidated text of the International Convention for Safety of Life at Sea, 1974, and its Protocol of 1988: articles, annexes and certificates. Incorporating all amendments up to and including the 1990 amendments (1991 amendments as an appendix),* International Maritime Organization, London, United Kingdom, 1992. ISBN 978-92-801-1594-5

11. SOLAS Consolidated Edition, 2001 *Consolidated text of the International Convention for Safety of Life at Sea, 1974, and its Protocol of 1988: articles, annexes and certificates. Incorporating all amendments in effect from 1 January 2001,* International Maritime Organization, London, United Kingdom, 2001. ISBN 92-801-5100-2

12. SOLAS Consolidated Edition, 2004 *Consolidated text of the International Convention for Safety of Life at Sea, 1974, and its Protocol of 1988: articles, annexes and certificates. Incorporating all amendments in effect from 1 July 2004,* International Maritime Organization, London, United Kingdom, 2004. ISBN 92-801-4183-X

13. SOLAS Amendments 2003, 2004 and 2005, International Maritime Organization, London, United Kingdom, 2006. ISBN 978-92-801-4224-2

14. SOLAS 2006 Amendments, International Maritime Organization, London, United Kingdom, 2007. ISBN 978-92-801-1492-9

15. WOODWARD, M.D. and HUTCHINSON, K.W. 'Uncertainty in ship stability', The Naval Architect, The Royal Institution of Naval Architects, London, United Kingdom, September 2016. ISSN: 0306-0209

16. Directive 2003/25/EC of the European Parliament and of the Council of 14 April 2003 on specific stability requirements for ro-ro passenger ships, Official Journal L123, of the European Union, Brussels, Belgium, 17th May, 2003.

17. Resolution A.265(VIII) *Regulations on Subdivision and Stability of Passenger Ships as an Equivalent of Part B of Chapter II of The International Convention for the Safety of Life at Sea, 1960*, Assembly: Eighth Session, 13th to 23rd November 1973, Resolutions and Other Decisions. Inter-Governmental Maritime Consultative Organization, London, United Kingdom, 1974.

18. SOLAS Consolidated Edition, 2009, *Consolidated text of the International Convention for Safety of Life at Sea, 1974, and its Protocol of 1988: articles, annexes and certificates. Incorporating all amendments in effect from 1 July 2009,* International Maritime Organization, London, United Kingdom, 2009. ISBN 978-92-801-1505-5 (note Erratum for Part B-1: Stability Regulation 7-2 (3) on page 47 re s_{final})

19. SOLAS Consolidated Edition, 2014, *Consolidated text of the International Convention for Safety of Life at Sea, 1974, and its Protocol of 1988: articles, annexes and certificates. Incorporating all amendments in effect from 1 July 2014,* International Maritime Organization, London, United Kingdom, 2014. ISBN 978-92-801-1594-9

20. SLF 55/17 *Report to the Maritime Safety Committee*, 55th Session of the Sub-Committee on Stability and Load Lines and on Fishing Vessels Safety, International Maritime Organization, London, United Kingdom, 25th February, 2013.

21. SDC 3/21 *Report to the Maritime Safety Committee*, 3rd Session of the Sub-Committee on Ship Design and Construction, International Maritime Organization, London, United Kingdom, 23rd February, 2016.

22. MSC 96/11 Ship Design and Construction: *Report of the third session of the Sub-committee: Note by the Secretariat,* International Maritime Organization, London, United Kingdom, 24th February, 2016.

23 *Study Assessing the Acceptable and Practicable risk level of Passenger Ships related to Damage Stability* summarised in SDC 3/3/7 (Proposals for the Required Index, *R*) including a link to the full reports on the EMSA web-site.

9. AUTHORS BIOGRAPHIES

Keith W Hutchinson is Engineering Manager and Chief Engineer for Commercial Marine with Babcock Energy and Marine at their Tyneside offices in North Shields, United Kingdom, where he undertakes commercial and offshore marine consultancy for a range of clients globally. He is a Ship Designer and Naval Architect, and is responsible for the coordination and management of all engineering teams and, as senior Technical Authority, also for all governance relating to whole ship design, engineering etc. He is Subject Matter Expert (SME) for the Naval Architecture domain across all Babcock offices. Keith leads multidisciplinary teams formulating design solutions, together with cost estimating and investment appraisals. Over a thirty year career he has been constantly at the forefront of the technical development of robust but flexible concept / basic designs for a multitude of extremely novel / complex / bespoke merchant, offshore and naval vessels, both new-build and conversions – ranging from sub 20 metre high speed catamarans, landing craft etc. to FPSOs and FLNGs in excess of 500 metres long / 700,000 tonnes displacement for the harshest environments, consistently striving with clients to explore unknown and commonly conflicting multiple dimensional design spaces.

Keith graduated in Naval Architecture and Shipbuilding from Sunderland Polytechnic in 1990 and from the University of Newcastle upon Tyne in 1992 with a Bachelor of Engineering in Marine Technology with First Class Honours in Naval Architecture. Having worked in several Shipbuilding and Shiprepair yards on the River Tyne, he joined Swan Hunter Shipbuilders in 1990 and in 1992 was appointed Senior Designer within the Naval Architecture Section of the Basic Design Department. Upon joining his current employer in 1995, then Armstrong Technology, he was seconded to the Newcastle University's Engineering Design Centre (EDC) for over three years as a Research Associate developing decision support tools for the evaluation / optimisation / selection of safer RO-RO ship designs etc.

Keith has published over 40 technical papers and articles on naval architecture and ship design etc. in addition to being invited to give almost 30 presentations on aspects of ship / offshore vessel design and performance. He is a Member of Lloyd's Register of Shipping's (LRS) main commercial ship Technical Committee (TC), Offshore Technical Committee (OTC) and Naval Ship Technical Committee (NSTC). He is a Member of SNAME's Ship Design (SDC) and Offshore Committees (OC), and various Panels of these. He is a Chartered Engineer (CEng), Chartered Marine Engineer (CMarEng), Fellow of The Royal Institution of Naval Architects (FRINA), Fellow of The Institute Marine Engineering, Science and Technology (FIMarEST) and Fellow of The Society of Naval Architects and Marine Engineers (FSNAME). He has been active in the North-East Coast Joint Branch (NECJB) of IMarEST and RINA for over two decades, serving as Honorary Treasurer for nearly fifteen years and previously as Chairman and Vice-Chairman, and a Committee Member. He is a founding Committee Member and Technical and Research (T&R) / Paper Chair of SNAME's Western Europe Section (WES).

Andrew L Scott is Policy Lead for Stability, Load Line and Tonnage within the Marine Technology Branch of the Maritime Safety and Standards Directorate of the Maritime and Coastguard Agency (MCA) based in their Tyne offices in South Shields, United Kingdom. He is responsible for policy matters relating to stability, load line and tonnage and was, until recently, co-ordinator of the International Maritime Organization's (IMO) Ship Design and Construction (SDC) Correspondence Group covering the latest updates to the Safety of Life at Sea (SOLAS 2009) and on issues relating to Roll-On / Roll-Off (RO-RO) ship stability. He has led several MCA-sponsored Research Programmes into the impact of the 2009 SOLAS amendments on ship design most specifically on the issue of Water-on-Deck (WoD, the Stockholm Agreement).

Andrew graduated from the University of Newcastle upon Tyne in 1969 with a BSc degree in Naval Architecture and Shipbuilding and joined Lloyds Register of Shipping (LRS) as a trainee Ship Surveyor, spending seven years in their London HQ in various departments with a spell of three years working in Amsterdam on survey work, principally on Very Large Crude Carriers (VLCCs). Andrew then joined the British Ship Research Association (BSRA) (later BMT) in Wallsend, as a Research Officer working on a wide range of consultancy projects, particularly with RO-PAX ships, and on software development and support. Andrew has published joint papers on 'Ship Vulnerability to Flooding' (IMC on Design and Safety, Berkeley, 2007), 'Consideration of Damage to Ships from Conceptual Design to Operation' (RINA The Damaged Ship, 2011) and 'The Development and Impact of Current and Possible Future National and International Damage Stability Regulations (RINA The Damaged Ship II, 2013), 'Current and possible future Intact and Damage Stability Regulations and Verification Tools for Passenger Ships' (RINA The Damaged Ship III 2015), 'Passenger RO-RO Ferry Damage Stability – Status and Development of International Regulations' (RINA Design and Operation of Ferries and RO-PAX Vessels), and given a number of presentations on related topics, most recently at the Stability 2010 Workshop at Wageningen and at the GOALDS damage stability workshop in Oslo (October 2011).

Andrew is a Chartered Engineer (CEng) and a Member of The Royal Institution of Naval Architects (MRINA). He was appointed Member of the Most Excellent Order of the British Empire (MBE) in the 2016 Queen's Birthday Honours for his work helping to implement safety-critical improvements to international damage stability regulations.

DESIGNED ACCORDING TO,-OR FAR BEYOND, THE REGULATIONS?

R.C. Imstøl, Western Norway University of Applied Sciences, Norway

SUMMARY

The engineering slogan 'according to the rule', or rather 'bending the rule until the rules objective seems obscure', has recently been brought to the headlines by VW. In Bundeskansler Andrea Merkel words: *The way things were swept under the rug or where loopholes in emission tests were massively exploited to the point they were unrecognizable, that destroys trust* [1].
In this paper, the Emma Mærsk thruster incident, the Norman Atlantic burnout and the Al Salam Boccaccio 98 disaster as well as DUE DILLIGENCE are being discussed to address the responsibility of maritime authorities.

1. INTRODUCTION

In 2004 MV Rocknes hit a shoal in the estuary of Bergen and capsized 18 seafarers perished. One of the reasons for the brutal outcome was the huge wing / double bottom tanks that are common for ore and stone carriers. The Norwegian Maritime Administration (NMA) pointed, in their report, that the arrangement was common but not necessarily in line with the double bottom requirement of IMO [2].

In 2013 Emma Märsk suffered a thruster blade collapse and one of the aft thruster tunnels got leak. The flooding was not restrained to the one compartment since 6 cable penetrating sealings did not have the quality required. The axel-tunnel between the propeller and the engine room located midship was consequently flooded. As the door separating the axel-tunnel from the engine-room was gas but not watertight the engine room was flooded as well.

Reading the accident report on Emma Märsk [3] the double bottom of Rocknes once again caught one's attention. -Might it be; that naval architects pay attention only to the collision requirement neglecting the double bottom? If so, should the double bottom requirement be baked into SOLAS damage stability regulation rather than being on its own? Furthermore; should the double bottom requirement be formed in computerising terms?

In December 2014 MV Norman Atlantic burned out on her route from Greece to Italy and some months later the MV Sorrento burned out as well. In both cases, the fires had originated in an open ro-ro space and spread throughout the deck and upwards. The impact of the heavy smoke on system performance was in both cases disastrous. According to the naval architects involved these ships was designed according to the rules. However, studying the rules one found that the notion 'open ro-ro compartment' is a mere definition. It is written nowhere that the prescriptive regulation in the following chapters does not apply. Consequently these ships should be handled by the exemption rule.

To study the approval documentation of Emma Märsk as well as an open ro-ro ferry; the Danish Maritime Administration (DMA) was requested to disclose all relevant documents for the MV Emma Mærsk and MV Oslofjord [4] [5]. NOTE: Oslofjord was chosen because Italian authorities did not respond to requests to Norman Atlantic and one knew Oslofjord to be an open ro-ro offshore ferry plying between Norway and Sweden. For Emma Märsk, DMA passed on the matter to American Bureau of Shipping (ABS) which is her Recognized Organization (RO). For Oslofjord DMA seemed positive but dragged their feet and little documentation has been revealed. For Emma Märsk, ABS seemed negligent until them after 6 months and several complaints to DMA wrote:

> ABS can confirm that they have passed the request for information to the Owners of the Vessel who have responded as follows:
> I refer to your communication to the Danish Maritime Authority of the 2nd March 2018 concerning the disclosure of documents regarding EMMA MÆRSK.
>
> "We can inform you that Maersk cannot consent to such disclosure.
>
> Maersk' position is based on the provisions in the Danish Act on Public Access (In Danish: "Offentlighedsloven") which explicitly excludes (i) confidential private interest and (ii) technical information and business sensitive information concerning operational and commercial matters from the right of public access." [6]

It is understandable that owners, ROs, as well as maritime authorities find documentation of likely unseaworthiness sensitive but it seems far from understandable that such unseaworthiness is a private matter. The Danish Maritime Investigation Bureau should have investigated the severe floating equilibrium but they were to detail oriented to recognize that a mere incident in a thruster tunnel seemingly might have turned into a complete loss of ship and cargo under somewhat different circumstances. It shall be interesting to see how Danish authorities will handle the upcoming complaint in this matter.

Another ship, which springs to one's mind when it comes to violation of regulation by her flag, is the Al Salam Boccaccio 98 (ASB98). She capsized in the Red Sea in 2006 and more than thousand seafarers lost their lives. In spite of what RINA S.p.A. (her RO) says their archives hold no damage stability documentation for these 6 sister vessels' redesign under Italian flag. However, contrary to the examples previously mentioned one has no reason to suspect such violation to be generic / systematic throughout the industry. Having said so, it far from obvious that such violation is unique.

2. THE DOUBLE BOTTOM PROBLEM

In their Rocknes report NMA discuss the double bottom regulations [2 $3.2]:

> *Chapter II-1, Regulation 12-1, first paragraph requires a double bottom extending from the collision bulkhead to the afterpeak bulkhead as far as this is practicable and compatible with the design and proper working of the ship.*

With this in mind the Danish Maritime Accident Investigation Bureau's (DMAIB) Emma Mærsk report caught attention [$3.6]:

> *The bulkhead between the engine room and the shaft tunnel, situated at frame no. 166, was designed, constructed, approved and relied on as a watertight bulkhead (figures 30 and 31 on page 30).*
> *This bulkhead had not been constructed to meet the ship's damage stability conditions, but rather to limit the needed capacity of the CO2 fire extinguishing plant. It was not compulsory according to SOLAS regulations.*

While stone-carriers as Rocknes may have reasons to have huge double bottom tanks there may hardly be any reason not to make the door through this bulkhead watertight. The only reason one can think of is that one is not taking the double bottom requirement seriously. May it be that the authorities restrain themselves to verify the side damage regulations? If so, the industry and its seafarers may expect many unwanted loss of ship and life.

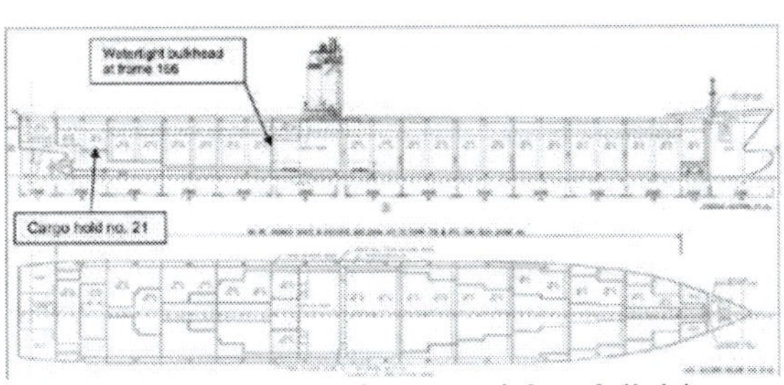

Figure 1: Emma Mærsk's watertight subdivision

2.1 CALL FOR NEW REGUALTIONS

If the present double bottom requirement may not be properly read by the IMO members one should add computer friendly rules to the Damage Stability Rules governing the double bottom challenge. Alternatively the Damage Stability Rules rules should be renamed Collision Stability Rules and the double bottom rules should be renamed Bottom Damage Stability Rules in order to clarify the matter.

2.2 DISCLOSURE OF DOCUMENTS FOR MV EMMA MÆRSK

Since the approval by DMA seems in statutory all relevant documents for Emma Mærsk was requested disclosed. However, as the ship was approved by American Bureau of Shipping (ABS) on behalf of DMA the request was further sent to ABS by DMA.

ABS seem far from eager to respond but after several complaints to DMA and a sharp notice from them to ABS the latter concluded that it was in the owners interest not to disclose any documentation. After another complaint to DMA they pointed to ABS that ABS had apparently not been in contact with the owners as the RO-contract required. More time passed, nothing happened and a new complaint was sent to DMA who sent a reminder to ABS.

Seven months after the Emma Mærsk documents was requested disclosed there is no reason to believe that the owner has been informed by neither DMA nor ABS that the double bottoms of 8 of their containerships may be in statutory.

The owners may however, have been informed by their in house insurers, Mærsk Insurance AS. The latter was at least duly informed by the author.

3. OPEN CARGO SPACE FIRES

In the aftermath of the Norman Atlantic and the MV Sorrento fires the Italian judge prosecutor nominated a committee to write a report on the accident [7]. The report is comprehensive but it does not discuss the role of the authorities and the present regulations or the industry's application of them. EMSA ordered another report, which has the same shortcomings as the NA-committee's report [8]. However, the EMSA report discusses open and closed ro-ro spaces in generic terms.

The author of this paper analysed above mentioned reports and chose to draft another report adding simple nautical knowledge. The draft was then sent to DMA, as an argument for disclosure of relevant documents for MV Oslofjord. DMA put a Cand. Polit. on the case and she promised to have DMAs experts commenting my report. Nearly 6 months later one still awaits DMA's comments in order to complete the draft report. The main part of this paper is based on the draft report.

3.1 THE EMSA REPORT

FIRESAFE carried out simulations of generic fires at generic open and closed ro-ro compartments. These simulations demonstrate the problem clearly:

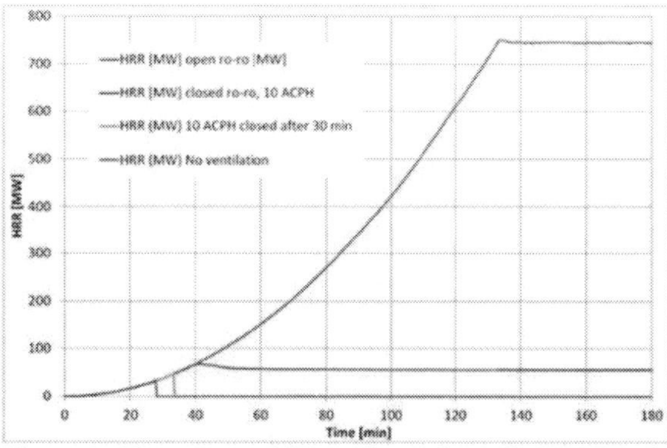

Figure 2: The expected heat release from fire on generic ro-ro decks

The blue graph shows the amount of energy fire a typical at an open ro-ro compartment will release as function of time. We see from the graph that aprx 37minutes after ignition the release will be aprx 30.000kWh (sufficient for 10.000 homes). 120 minutes after ignition the release will be aprx 600.000kWh (sufficient for 200.000 homes). 150 minutes after ignition the output will be aprx 750.000kWh (sufficient for 250.000 homes). Thereafter the output will be steady at this level until the deck is completely burned out. The open ro-ro deck arrangement works consequently as an uncontrollable furnace.

The cyanide graph shows the situation if the ventilators of a closed ro-ro compartment of similar make are closed immediately after ignition. NOTE: All the graphs are identical for the first 37 minutes but then this graphs drops when the heat release stops due to lack of air. The closed ro-ro deck arrangement works consequently like a candle-snuffer.

The green graph shows the situation for a closed ro-ro compartment of similar make if one leave the ventilation running for 30 minutes.

The red graph shows the situation for a closed ro-ro deck if one leaves the ventilation running.

The red versus the blue graphs consequently shows what the fire-extinguishing system should be able to handle for closed versus open ro-ro compartments. It is in fact much worse since the fire-extinguishing system becomes the only form for fire-control system for open compartments while it is only an add-on for closed compartments. Consequently, a fire-extinguishing system, which may handle an open compartment fire, must be more powerful and it must have at least one fully redundant system.

3.2 RELEVANT SOLAS REGULATIONS

SOLAS CHAPTER II-2, Construction – Fire Protection, fire detection and fire extinction has 7 parts. The first of these parts (A) is general and it is particularly pointed in regulation 2.3 that one has to comply to the other parts, the prescriptive parts (B, C, D, E & G) or to the exemption part (E).

The most important prescriptive regulation is, as seen from the EMSA graph, regulation 5 which indeed says *means of control for the air-supply to the space shall be provided.* Since this regulation obviously was not satisfied in the MV Norman Atlantic and the MV Sorrento, cases; NMA was questioned which regulations governed the open ro-ro spaces. Regulation A.3.35&36 was the answer. The problem is; this is a definition not a prescriptive regulation. Rather than applying *Part F – Alternative design and arrangements* one seems to have pretended that the prescriptive regulations suitable for closed ro-ro deck might be used without any further analyses.

Figure 3: MV Sorrento after burnout.

There is a bunch of other prescriptive regulations which as well should be considered as alternative under Part F. It should not be necessary to go into detail here. One may merely conclude that members of IMO seem collectively unable to read SOLAS and consequently seafarers suffers such burn-outs as seen on figure 3;

3.3 DISCLOSURE OF DOCUMENTS FOR MV OSLOFJORD

Since Italian authorities did not respond to several requests the author chose to request disclosure of documents for an open ro-ro ferry plying between Norway and Sweden. MV Oslofjord was originally built under Norwegian flag. However, by commercial reasons she was reflagged to Denmark where she later has been rebuilt. She is consequently not an open ro-ro ferry any more.

This ship has nearly fully been approved by DMA thus there has not been any other party confusing DMA's responsibilities. This request for disclosure has been running back and forth for more than 6 months and DMA seems as reluctant to disclose documents for Oslofjord as ABS for Emma Mærsk.

4. AL SALAM BOCCACCIO 98

Figure 4: ASB 98 before conversion.

Preparing the litigation against the Italian classification association RINA S.p.A. one first requested copies of all relevant documents. That did not work out so one had to bring the issue in for Italian court. RINA S.p.A. was thus sentenced to hand out all the document for the time ASB 98 and her sister a vessel was sailing under Italian flag. Unfortunately, these documents do not include any damage stability calculi approved by RINA S.p.A. for the redesign of these ships. Even more unfortunate, Damage Stability Calculi carried out by the software provider used by RINA S.p.A. at the time of these ships redesign conclude that these ships would not have sufficient stability if the sternmost compartment had been damaged.

Figure 5: ASB 98 after conversion.

Whether RINA S.p.A. approved any damage stability calculi when these ships was reflagged to Panama. Apparently, the Egyptian investigator was not at all satisfied.

5. COLREG, RULE §2, RESPONSIBILITY

As a merchant navy officer student I had to study the International Regulations for Preventing Collisions at Sea. Particular focus was put on the responsibility rule, which we had to learn, by heart. It states:

> *(a). Nothing in these Rules shall exonerate any vessel, or the owner, master or crew thereof, from the consequences of any neglect to comply with these Rules or of the neglect of any precaution which may be required by the ordinary practice of seamen, or by the special circumstances of the case.*
>
> *(b). In construing and complying with these Rules due regard shall be had to all dangers of navigation and collision and to any special circumstances, including the limitations of the vessels involved, which may make a departure from these Rules necessary to avoid immediate danger.*

SOLAS has something similar: SOLAS, Part I, Article I (b) says: *The Contracting Governments undertake to promulgate all laws, decrees, orders, and regulations and to take all other steps which may be necessary to give the present Convention full and complete effect so as to ensure that, from the point of view of safety of life, a ship is fit for the service for which it is intended.*

This is apparently far too weak and modified COLREG Rule §2 seem called for:

> (c1) Nothing in this Convention shall exonerate any; Flag, or Recognized Organization, or owner, or shipyard, or designer, or insurer, from the consequences of any neglect to comply with these Rules, or of the neglect of any precaution from the point of view of safety of life, to ensure that a ship is fit for the service for which it is intended.
> (c2) In construing and complying with this Convention due regard shall be had to all dangers of navigation and to any special circumstances, including the limitations of the vessel in question, which may make a deviation of SOLAS necessary to avoid risk of life at sea.

6. DUE DILLIGENCE

Studies of ship incidents, accidents and disasters such as MV Rocknes, MV Estonia, MV Al Salam Boccaccio 98, MV Saint Malo, MV Norman Atlantic, MV Sorrento, MV Emma Mærsk suggest that the maritime industry and as well as its governors has understanding of the juridical notion DUE DILLIGENCE: *The care that a reasonable person exercises to avoid harm to other persons or their property* [9].

7. CONCLUSIONS

1. Emma Mærsk double bottom seems very vulnerable for bottom scratches involving the axel tunnel.
   ---
2. Norman Atlantic's design deviated strongly from international rules and principles of fire-control.
   a. Her design did however, not deviate from international design custom.
3. Due to her design and in case on a fire on her open ro-ro space (Deck #4):

a. Her fixed fire extinguishing system must, throughout her life, be considered neither functional nor statutory.
b. All of her fire pumps must, throughout her life, be considered neither functional nor statutory.
c. All of her evacuation systems must, throughout her life, be considered neither functional nor statutory.
d. All of her control stations must, throughout her life, be considered neither functional nor statutory.
e. All of her main engines, generators, and emergency generators must, throughout her life, be considered neither functional nor statutory.
f. All of her fire detectors at deck #4 must, throughout her life, be considered neither functional nor statutory.

---

4. As all above seems generic for offshore ships with open ro-ro spaces it is unreasonable to assume that any of these ships, at any time, from keel to grave, may have had a sea-worthy voyage.

---

5. Al Salam Boccaccio 98 did not have approved damage stability when she and her sister vessels were converted.
6. In the aftermath of the Al Salam Boccaccio 98 approved damage stability has not been documented.

8. AUTHORS BIOGRAPHY

Rolf C. Imstøl holds the position of assistant professor at Western Norway University of Applied Sciences. He is a former merchant navy officer who has lectured nautics, naval architecture and computation. He turned a whistle blower in Rocknes case demonstrating that the ship's loading calculator had been manipulated by use of too heavy bulk density. He did as well prove that the Estonia *conspiracy* hypotheses heralded by the ship builders experts is the only hypothesis that makes any physical sense. Later he has been assisting lawyers assisting bereaved, survivors and other raising their claims against the industry.

9. REFERENCES

[1] Merkel at CDU party, Aug 2017, https://www.pressreader.com/kuwait/arab-times/20170813/282003262527454, 2017

[2] NMA, Report from the working group charged with considering the ship-technical aspects of the capsizing of Rocknes – V2PU3 – Antigua and Barbuda,19 January 2004 in Vatlestraumen off Bergen, Norway, 2004

[3] DMAIB, EMMA MÆRSK Flooding of engine room on 1 February 2013, DMAIB #2013002165, 2013

[4] DMA, Access to documents for Emma Mærsk, DMA # 2018006707, 2017-2018

[5] DMA, Request for disclosure of documents due to the Norman Atlantic and Sorrento burnouts, DMA #2017040665, 2017-2018

[6] Niels Bruun, American Bureau of Shipping, Access to documents for Emma Mærsk, DMA # 2018006707, 24.04.2017, Copenhagen

[7] Technical report by the consultants nominated by the GIP, Penal case N° 20598/2014 RGNR 489/15 RGIP. 2017

[8] FIRESAFE , Study investigating cost effective measures for reducing the risk from fires on ro-ro passenger ships, (FIRESAFE)EMSA report, EMSA/OP/01/2016

[9] Merriam-Webster, https://www.merriam-webster.com/dictionary/due%20diligence, 2018

ASSESSMENT OF RAKING DAMAGE FROM DECELERATIONS MEASURED DURING A GROUNDING INCIDENT

E.C. van Kempen, MSc student, TU Delft, The Netherlands
A.W. Vredeveldt, senior scientist, TNO, The Netherlands
M.G. Hoogeland, project manager, TNO, The Netherlands

SUMMARY

A three degree of freedom external dynamics calculation model for ship grounding has been made, which is used to predict the extent of bottom raking damage and the grounding force with associated decelerations of the ship in terms of surge, sway and yaw. When the ship decelerates, caused by a grounding event, these motions can be logged through continuous ship motion measurements. From these deceleration time traces it is possible to calculate the bottom damage path through a double integration with respect to time. Moreover, with these decelerations, the location of the raking damage along the ship's bottom can be calculated as well. A straight forward method is proposed, which is validated against data on ship grounding external dynamics from earlier research and data from large-scale grounding experiments, both carried out in the nineties of the previous century.

Keywords *grounding damage, prediction from measured accelerations, ship grounding, raking damage, three degrees of freedom external dynamics model, ship survival*

NOMENCLATURE

DoF	Degrees of freedom
CoG	Ship centre of gravity
CL	Centre line
FPOI	First point of impact
LCG	Longitudinal centre of gravity
PP	Perpendiculars

1. INTRODUCTION

When a grounding incident occurs, the officer of the watch is usually unable to assess the severity of the damage. Most ships do have damage response procedures in place. In case of military ships these are well developed, where a well-trained crew, in the form of damage parties, is available to respond adequately. For non-navy ships, this is hardly possible, mostly due to crew restrictions. But even for navy vessels, underwater damage is hard to assess because of accessibility issues of flooded compartments. This is undesirable because it needs to be known which compartments have flooded in order to assess the situation with respect to residual buoyancy and stability. Based on the outcome, the officer in charge can decide whether to evacuate or to stay on board and control the situation. For establishing which compartments will flood the extent of a raking damage must be known. It is proposed that the accelerations (or rather, decelerations) measured on board during a grounding incident can be used to make a prediction of the extent of raking damage. From the deceleration time-traces, the stopping distance of the ship can be calculated. Subsequently, in the case of a grounding on a rigid rock, this stopping distance could be used as a measure of the raking damage length. A study is performed in order to investigate the viability of this idea. An analysis of accelerations measured during large-scale grounding experiments [1] shows that the use of deceleration time traces can proof useful in the prediction of the raking damage extent. This is illustrated in Figure 1. In the figure the bottom view of the damaged test section after the grounding experiment (three holes and deformation path are visible) and the rock are shown along with the absorbed energy calculated from the horizontal grounding force and the horizontal rock penetration, the energy calculated based on the surge motion measured at midship and finally the ship's pitch angle. From the analysis of the acceleration time traces it is found that the raking damage length, rupture initiation, sliding and floor passages (i.e. rock striking an internal floor) can be observed from them.

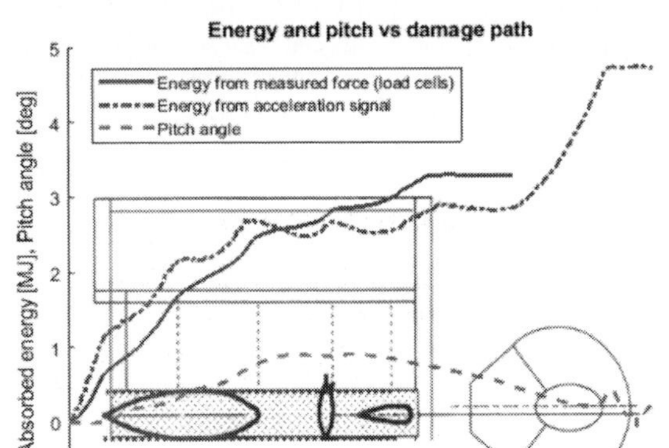

Figure 1: Absorption of kinetic energy and pitch angle, plotted on bottom view of the damaged test section [1]

The method described in this paper is intended as an aid on the bridge for the assessment of the severity of the situation after a grounding incident, without the use of damage reports from the crew. The method does not rely on elaborate internal mechanics calculations, which makes detailed knowledge of the ship structure necessary. Instead, the proposed method makes use of the decelerations measured on board during a grounding incident. Measuring ship motions with the use of accelerometers is not expensive. In the most simplest of ways, the method could already be applied with the use of a mobile phone, registering these decelerations, albeit that measurement accuracy with such a device may prove insufficient.

In the study a three degrees of freedom (DoF) external dynamics grounding model (surge, sway and yaw motions) has been implemented. This calculation model is validated against a sophisticated model of Simonsen [2] and against the TNO-ASIS large-scale test data [1]. After that, a limited range of grounding scenarios were simulated, yielding deceleration time traces. Finally, solely based on these deceleration, time traces the rock-bottom contact path is predicted, resulting in the extent of the raking damage.

2. THREE DEGREES OF FREEDOM GROUNDING MODEL

2.1 EQUATION OF MOTION

The three degrees of freedom of the ship motions considered are surge, sway and yaw. These motions are depicted by respectively u, v and ψ. The local coordinate system, xyz, is fixed to the ship's centre of gravity (CoG). The relative distance between the rock and the ship's CoG is (X_s, Y_s). A representation of the grounding model's ship coordinate system is given in Figure 2. Here the ship is first shown just before grounding. Then it is shown after the ship hits the rock off-centre (distance from ship centre line (CL) is μ). This results in a damage path depending on the first point of impact (FPOI). The deviation from a straight damage path at time t is δ. The equation of motion used in the three DoF external dynamics model is given in Equation 1.

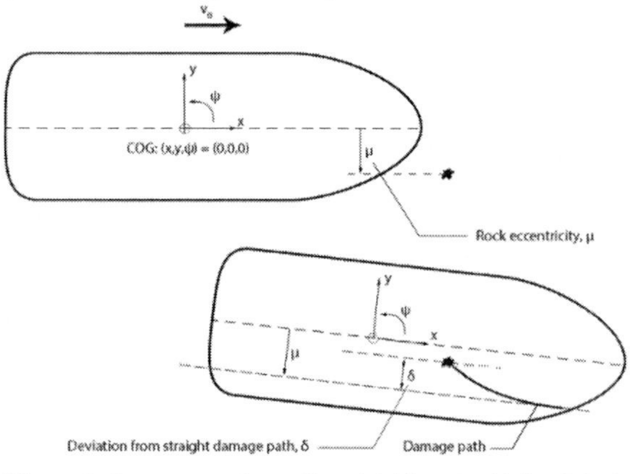

Figure 2: Representation of local (ship's CoG) in global (rock) coordinate system (based on Figure 7.1 in [2])

$$\begin{bmatrix} M+A_{11} & 0 & 0 \\ 0 & M+A_{22} & A_{26} \\ 0 & A_{62} & I_6+A_{66} \end{bmatrix} \begin{Bmatrix} \ddot{u} \\ \ddot{v} \\ \ddot{\psi} \end{Bmatrix} = \begin{Bmatrix} F_x \\ F_y \\ M_z \end{Bmatrix}_G + \begin{Bmatrix} M_{22}(\dot{v}\dot{\psi} + x_{gy}\dot{\psi}^2) \\ -M_{11}\dot{u}\dot{\psi} \\ (M_{11}-M_{22})\dot{u}\dot{v} + x_{gy}M_{22}\dot{u}\dot{\psi} \end{Bmatrix} \quad (1)$$

The first matrix on the left side in Equation 1 is the mass + added mass matrix. $M_{11} = M + A_{11}$ and $M_{22} = M + A_{22}$ denotes respectively the ship mass plus added mass [kg] in surge and sway. $M_{66} = I_6 + A_{66}$ denotes the moment of inertia and added mass moment of inertia about the ship's vertical-axis. The sway and yaw motions are coupled using the term $x_{gy}M_{22}$ [kg m]. Due to the ship's lateral symmetry, the surge motion is decoupled and the added mass terms A_{26} and A_{62} are equal.

The first vector in the right-hand side of the motion equation is the grounding force due to the impact with the rigid rock.

The second vector in the right-hand side of the equation of motion is applied to take into account the Coriolis-centripetal forces. These fictitious forces must be applied due to choice of frame of reference and are treated as external loads. When the heave, roll and pitch motions are neglected, then the Coriolis-centripetal term is as presented in the vector in Equation 1. In this simulation, these forces are only present when there is a change of yaw angle.

During the simulation, the motion equation is solved through a integration in the time domain and the accelerations of the ship using the initial ship velocity are calculated. From this, the velocity and position of the ship as a function of time are calculated.

2.2 VALIDATION AGAINST LARGE-SCALE EXPERIMENT

The model is first validated against the TNO-ASIS large-scale ship grounding experiment. In this experiment, a rock offset of 0.3 m was observed and is taken into account in the simulation. The length of the measured damage path is the full length of the test section, i.e. 5.5 m. At this point 3.4 MJ is dissipated (based on load cells and wire gauges, Figure 1). This implies an average longitudinal grounding force on the ship of 0.62 MN.

The yaw angle at termination of the simulation of this scenario, ψ_{term}, is -0.21 deg. The deviation from a straight damage path, δ, is 0.06 m. The offset of 0.3 m can therefore be neglected and the simulation can be viewed as a one DoF problem.

A factor that influences the simulation's result is the inertia. A good way of determining how well the first estimate of the inertia of the large-scale experiment is, is by looking at the absorbed energy. From Figure 1, it can be observed that the absorbed energy calculated using the acceleration signal and the absorbed energy calculated from the measured force show the same trend. The energy from the acceleration signal is calculated by multiplying this signal by the ship's mass in surge and integrating along the damage path. The total surge mass in the experiment has been determined at 681 tonne (from the experiment's report). The line of the absorbed energy from acceleration shows an overestimation of the energy in the first half of the diagram (up to 3 m, Figure 1) and an underestimation in the second half. The variation of estimated ship mass contributing in the ship motion can be physically explained by the sloshing of the water in the tanks that is used to trim the ship to its initial trim angle before impact. Not all of the ballast water is initially 'felt' dynamically in the surge motion (leading to a difference in estimated damage length and absorbed energy), because of the effect of sloshing in the partially filled tanks. From the loading data of the experiment it is observed that 43% of the total weight is tank contents of any sort. It is not reported which tanks were partially filled however, but a maximum weight of 291 tonne could contribute to it. A minimum of 410 tonne total effective weight in the first part of the grounding and a maximum of 971 tonne effective weight in the second part of the grounding is how large the uncertainty in the weight estimation can be.

2.3 VALIDATION AGAINST SOPHISTICATED MODEL

The three DoF external dynamics model is further validated using the example of Simonsen [2]. In this example a 322 m VLCC is used. Using the data of this ship and Simonsen's results, the sensitivity of the model and the behavior of the ship motions have been investigated on the basis of varying the following parameters:

- Magnitude of grounding force, F_G;
- Initial velocity, v_0.

In order to validate the three DoF model, the results were checked with those of Simonsen. The result of the simulation are the damage path. This is expressed in, δ, the deviation from a straight damage path (depicted in Figure 2). Positive δ means deviation towards CL.

2.3 (a) Case 1: magnitude of grounding force

In the first case study the magnitude of the grounding force is varied. The rock eccentricity is taken almost at a maximum (almost half of the ship's beam), μ = 26 m. Figure 3 shows the damage path on the bottom of the ship for seven constant grounding forces. The grounding force, F_G, is calculated using the relation in Equation 2, where d_s is the desired stopping distance.

$$F_G = \frac{E_{kin,init}}{d_s} = \frac{\frac{1}{2}M_{11}v_0^2}{d_s} \qquad (2)$$

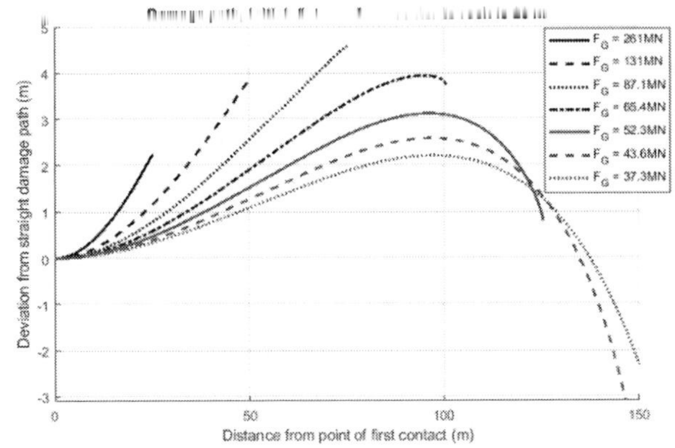

Figure 3: Damage path, varying grounding force

From the plot it can be seen that the lower the constant grounding force is taken, the longer the damage path is. Practically seen, the magnitude of the damage is smaller but the ship encounters not enough force to be stopped quickly. It can be seen that the ship locks onto the rock as it proceeds and turns towards it. This means that the damage path deviates from a straight damage path towards the ship's CL. As the damage path gets closer to the ship's CoG, the sway motion becomes larger and the ship turns away from the rock again. The damage path will turn towards the original path again and further towards the bilge. A more detailed view on the damage paths is presented in Table 1, where ΔX_S is the simulated damage length. The last two damage paths (F_G = 43.6 MN and 37.3 MN) leave the ship before stopping de simulation (out of bounds). Results are compared with [2] to validate the model and are presented in Table 1 as well, with [2] indicating Simonsen's results.

Table 1: Results case 1, comparison with results of [2]

F_G [MN]	$\frac{E_{kin,init}}{F_G}$ [m]	ΔX_S [m]	δ_{max} [m]	δ_{max} [2] [m]	ψ_{term} [deg]	ψ_{term} [2] [deg]
261	25	25.2	2.23	2.15	-1.83	-1.60
131	50	50.4	3.88	3.93	-4.09	-3.60
87.1	75	75.6	4.57	5.03	-7.11	-6.20
65.4	100	100.6	3.93	4.93	-10.87	-9.20
52.3	125	125.5	3.11	-	-15.23	-
43.6	150	150.0	o.o.b	2.49	-19.56	-15.70
37.3	175	174.7	o.o.b	-	-24.17	-

2.3 (b) Case 2: varying initial velocity

In the second case, the initial grounding velocity, v_0, is varied. The velocities are varied from: 4, 8, 12 and 16 kts (2.06, 4.12, 6.17 and 8.23 m/s). This is done at the grounding force where the δ was the biggest in Case 1. This is when the stopping distance is approximately 75 m, giving a grounding force of 87.1 MN and a deviation of the straight damage path of around 4.5 m. This is taken in order to get the results in a worst case scenario. For this study the initial rock eccentricity is chosen at 26 m. In Figure 4 the damage paths are plotted for each initial velocity.

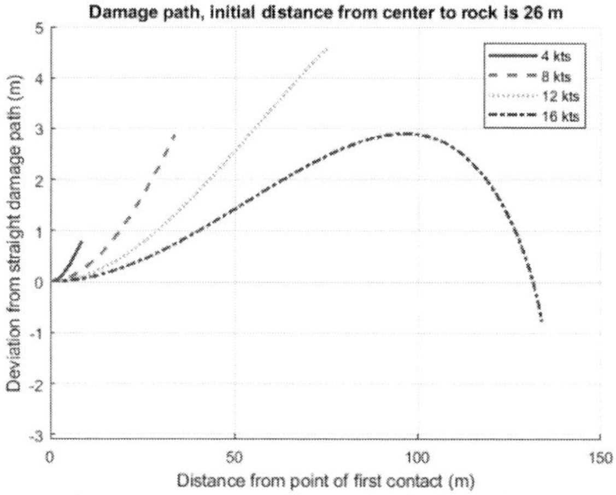

Figure 4: Damage path, varying initial velocity

The initial velocity changes the initial kinetic energy, thus being of direct influence on the damage length. From Figure 3 and 4 it can be seen that for longer damage lengths (end of damage path closer to the ship's CoG), the damage path bends back away from the CL again. In Table 2 the details of the simulation are given and results are compared.

Table 2: Results case 2, comparison with results of [2]

v_0 [kts]	$\frac{E_{kin,init}}{F_G}$ [m]	ΔX_S [m]	δ_{max} [m]	δ_{max} [2] [m]	ψ_{term} [deg]	ψ_{term} [2] [deg]
4	8.36	8.42	0.80	0.75	-0.55	-0.48
8	33.46	33.69	2.87	2.84	-2.48	-2.20
12	75.03	75.59	4.56	5.03	-7.10	-6.20
16	133.52	133.80	2.91	2.91	-16.67	-13.40

2.4 FURTHER SIMULATIONS OF THE GROUNDING MODEL

The three DoF external dynamics model is further investigated varying two more parameters:

- Rock eccentricity, μ;
- First point of impact (FPOI) in longitudinal direction, x_r.

2.4 (a) Case 3: varying rock eccentricity

The rock eccentricity is varied: 0, 6.5, 13, 19.5 and 26 m. This is done at an initial velocity of 12 kts and again the same constant grounding force as in case 2. The damage paths are plotted in Figure 5. The maximum deviation seems to increase proportionally to the increase in rock eccentricity.

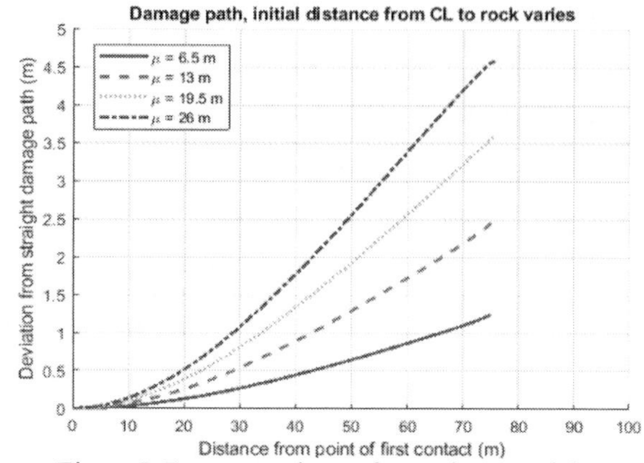

Figure 5: Damage path, varying rock eccentricity

2.4 (b) Case 4: Damage path, varying FPOI

Finally, the first point of impact (FPOI) is varied along the ship's longitudinal axis. The initial velocity is again 12 kts. The rock eccentricity is taken 13 m for all variations of longitudinal FPOI in this case 13 m is chosen because with this value the damage path has room to deviate the same distance both positively (towards the ship's CL) and negatively (towards the bilge). The main purpose of this study case is to look at the behavior of the damage path of a different first point of impact in longitudinal direction. Especially the behavior is interesting when the first point of impact is behind LCG. Of course, the majority of grounding scenarios occur in front of midship because of the ship's forward speed. But, for example, when the ship is trimmed, this grounding scenario is possible as well. In Figure 6 the damage paths are plotted. But this time they are plotted along the ship's longitudinal axis.

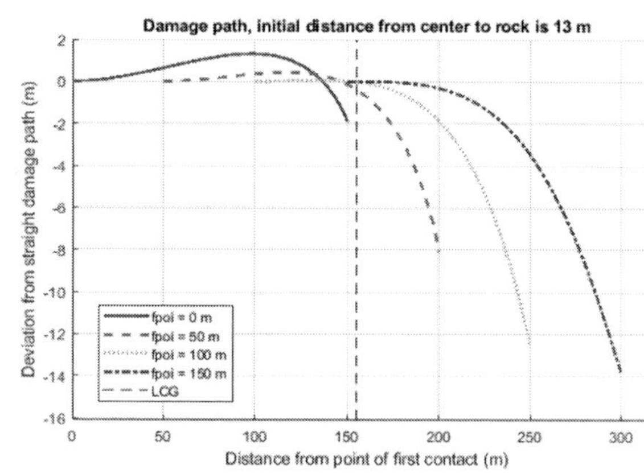

Figure 6: Damage path, varying FPOI (ΔX_S = 150 m)

From the plot it can be seen that the closer the FPOI lies to the CoG, the less the ship initially pivots towards the rock, which dampens the tendency of the damage path to deviate towards the CL. At FPOI = 150 m (almost at LCG), the damage path moves away from the CL.

2.5 OBSERVATIONS

From the figures 3 to 6, it is observed that the ship first locks on to the rock and the damage path first deviates towards the ship's CL. For longer damage paths, the damage path bends back to its original position and even tends to leave the ship. Also, as the damage path starts more towards LCG or behind it, the damage path tends to leave the ship right away.

For longer damage lengths (meaning smaller grounding forces), the results begin to differ more from the damage paths calculated by Simonsen. These differences can be explained by the omission of the damping term in Equation 1. When rock contact forces are small, the influence of the damping forces become relatively more important.

2.6 ASSUMPTIONS

In the case of a rock eccentricity of 26 m the ship hits the rock at the bilge, but the actual behavior of the ship would mean that at this far edge of the ship hull, the ship will tend to roll to the opposite side. But because we are looking at a three DoF model which only takes into account sway, surge and yaw, this is neglected. We are therefore looking at the worst case scenario: maximum yaw moment. Secondly, it is assumed that the ship has a constant beam between PP. This is easier in making clear a damage path has left the ship or not.

3. THREE CHARACTERISTIC GROUNDING SCENARIOS

In section 2 the damage paths are plotted for a grounding event with a constant grounding force, an initial velocity and a given first point of impact. In a real grounding scenario, all this data however is not known (except velocity). After an actual grounding, it is desired to know the estimated location of the resulting damage path. A method has been developed to investigate if it is possible to track down the rock contact path from just the accelerations in surge, sway and yaw measured on board.

The damage paths and acceleration time traces generated from the simulations for three FPOI locations were used to investigate this. The simulations are done with the three DoF external dynamics grounding model, while the 322 m VLCC example was used. The three locations in the ship's coordinate system at the start of the simulation are:

- $(x,\mu) = (50,0)$ m
- $(x,\mu) = (50,13)$ m
- $(x,\mu) = (180,13)$ m

The damage paths resulting from these impact locations are presented in Figure 7 (not to scale!). The second impact location shows only a small deviation from a straight damage path of less than ± 0.7 m. In the figure this damage path looks therefore almost straight. The third location shows a deviation from a straight damage path of about -7.5 m (towards bilge). Typical levels of accelerations observed are 200 mm/s^2 in surge and 20 mm/s^2 in sway.

3.1 CALCULATION OF ROCK CONTACT PATH

At this point, only the acceleration signals were looked at. All other data including the damage paths and the location of impact are forgotten. The proposed method must only use the decelerations and the initial velocity and should be able to make a prediction of the original impact location and rock contact path. The method is referred to as the inverse method.

The acceleration signal can be integrated twice in order to get the local ship velocity and displacement, just as is done in the three DoF model in order to determine the damage path. Here the initial velocity is used as additional information, but this is generally known on board (the goal was to only use the accelerations of course). For the model it was assumed that the initial velocity before grounding is purely in longitudinal direction. It is also assumed that the position of the ship's CoG is known (at least the position of the LCG). Then, the equation of motion is used to calculate the grounding force in the ship's coordinate system. These local forces, F_X and F_Y, are transformed to forces in the rock coordinate system, $F_{G,X}$ and $F_{G,Y}$. From this the constant grounding force, F_G, is found. It is assumed that the grounding force resultant points in the direction opposite to the relative velocity between the ship and the rock. This relation can be used to calculate the distance between the ship's CoG and rock in x- and y-direction: X_S and Y_S. This assumption implies that the global grounding force in x- and y-direction has the same ratio as the relative velocity between ship and rock in x- and y-direction. The X_S and Y_S parameters describe at any

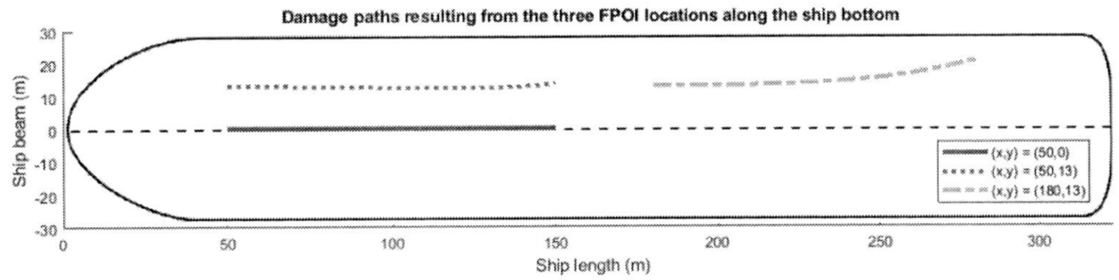

Figure 7: Damage paths along the ship bottom (top view)

moment in the simulation the position of the ship's CoG relative to the rock. At the start of the simulation (start grounding), these values represent the original position of the rock (FPOI). The X_S and Y_S coordinates calculated in each time step describe the rock bottom contact path.

3.2 RESULTS

The method is tested by calculating the first point of rock impact and the damage path. The results from the method were compared to the original rock contact path. An initial velocity of 12 kts is taken and a force is used so that the desired damage length is 100 m.

The result of the inverse method cannot find the FPOI location in case of impact at the CL. The result of the inverse method do equal the input data of the model in the cases of the off-CL location. But this is quite obvious for the same data is used for the inverse method as it is used to generate the acceleration time traces. Because it is assumed to ignore the rest of the data, an exact determination of the equation of motion is not possible. Next, the sensitivity of the model is tested.

3.3 OBSERVATIONS

For every test location the FPOI is predicted well, except for the situation where the ship hits the rock exactly at CL. This way the method has no way in knowing how large the forces in x- and y-direction are compared to each other (because the sway force and the yaw moment are zero). Because of this, the method cannot compute the arm of the yaw-moment and hence where along the ship's CL the grounding force acts. However, in a real situation an on-centre impact rarely happens. Also, including more data like the pitch could help determine the on-centre impact location.

4. SENSITIVITY ANALYSIS

The sensitivity of the method to ship inertias variation is tested by varying the inertia for the case of the damage path starting at (x,μ) = (50,13) m. As mentioned before, the method strongly relies on the estimation of the mass/moment of inertia and added mass of the ship. The masses and inertia are varied by a fixed amount. The mass in surge and sway and the moment of inertia are varied ± 20% and ± 40% of their original values.

4.1. RESULTS

The results of the inverse method using these variations are presented in Table 3 and Table 4. Here every column shows a variation, first of the mass in surge, then the mass in sway and finally the moment of inertia about the vertical axis (yaw). The coordinates of the FPOI result is given in the 4th and 6th row. The deviation from the original FPOI is given in the 5th and 7th row.

Table 3: FPOI prediction with ± 20% mass variation

M_{11} (kg)	-20%/+20%	$M_{11,0}$	$M_{11,0}$
M_{22} (kg)	$M_{22,0}$	-20%/+20%	$M_{22,0}$
M_{66} (kgm²)	$M_{66,0}$	$M_{66,0}$	-20%/+20%
x (m)	50.95/49.59	49.89/50.18	49.54/50.46
Error x	+1.9%/-0.8%	-0.2%/+0.4%	-0.9%/+0.9%
μ (m)	15.51/11.12	12.43/13.52	11.4/14.6
Error μ	+19.3%/-14.5%	-4.4%/+4.0%	-12.3%/+12.3%

Table 4: FPOI prediction with ± 40% mass variation

M_{11} (kg)	-40%/+40%	$M_{11,0}$	$M_{11,0}$
M_{22} (kg)	$M_{22,0}$	-40%/+40%	$M_{22,0}$
M_{66} (kgm²)	$M_{66,0}$	$M_{66,0}$	-40%/+40%
x (m)	53.11/49.44	49.88/50.40	49.06/50.94
Error x	+6.2%/-1.1%	-0.2%/+0.8%	-1.9%/+1.9%
μ (m)	18.82/9.70	11.81/13.97	9.77/16.23
Error μ	+44.8%/-25.4%	-9.2%/+7.5%	-24.9%/+24.9%

From Table 3 and Table 4 it can be seen that varying the values of the inertias has a substantial influence on the damage path. However, if the ratio between the inertias remains fixed, the damage path is not affected. Apparently, the method relies on the ratio of the accelerations/forces.

4.2 OBSERVATIONS

From the sensitivity study it is observed that the larger the variation in one variable, the larger the deviation from the original damage path gets. The variation of the inertias have the largest influence on the damage path in transverse direction. The FPOI in this direction varies between +5.8 m and -3.3 m, while the FPOI in longitudinal direction only varies between +3.1 m and -1.0 m.

It can be concluded that the method very much relies on the ratio between accelerations in surge, sway and yaw, rather than on the actual values.

5. DISCUSSION AND CONCLUSION

It is noted that typical acceleration levels determined on the VLCC do not exceed 200 mm/s² (0.02 g). This implies the need for accelerometers with sufficient sensitivity. Moreover, because of the long-time duration of the observed signals, the accelerometers must be able to deal with the static accelerations as well.

From the sensitivity study of the method predicting the rock-bottom contact path, it is seen that the inertia ratios have a large influence on the rock contact path in transverse direction. Also, the ratio between the measured decelerations have a large effect on the predicted damage path.

A three degree of freedom external dynamics grounding model has been implemented and works satisfactorily. With the use of the model and the inverse method, the decelerations can be used to calculate damage paths.

From the validation it is observed that sloshing may be an important factor in the use of the external dynamics grounding model. Sloshing causes the tank contents to behave partly as an independent dynamic system. This probably needs to be taken into account in case of VLCC's and LNG carriers.

6. REFERENCES

1. VREDEVELDT, A.W. & WEVERS, L.T., 'Full scale grounding experiments', *Conference on prediction methodology of Tanker Structural Failure & Consequential Oil Spill*, 1995.

2. SIMONSEN, B.C., 'Mechanics of ship grounding', *PhD thesis*, 1997.

7. AUTHORS BIOGRAPHY

Elizabeth C. van Kempen is MSc student Marine Technology at the Delft University of Technology. She holds the current position of graduate student at the Netherlands Organisation for Applied Sciences (TNO).

Alex W. Vredeveldt graduated in 1986 at the Technische Hogeschool Delft as a Naval Architect. He started his career at Wilton Fijenoord Dockyards in Schiedam where he worked in the project office, dealing with ship conversions and the building of submarines. He then joined the Netherlands Organisation for Applied Sciences (TNO). There he has worked on a wide variety of topics, ranging from ship collision and grounding analyses, including full scale tests, to passenger evacuation analyses (and tests). His current position is senior scientists at TNO, where he is active on safety and resilience of ships and offshore structures, which includes topics such as LNG and hydrogen as bunker fuel, offshore wind turbines and large civil structures such as lock gates and bridges.

Martijn G. Hoogeland holds a master's degree in Naval Architecture from the Delft University of Technology. After having worked for Bluewater in Hoofddorp and SBM in Schiedam, he joined TNO's centre for Structural Dynamics. He is currently active as project manager.

A RISK PROFILE FOR ESCORTED TANKERS AND THEIR RESISTANCE TO COLLISION DAMAGE

I Braidwood, CEng, PhD, MRINA, MSNAME, Naval Architecture Consulting, Canada
R G Allan, P Eng, FEC, FRINA, FSNAME, Robert Allan Ltd., Canada

SUMMARY

Perhaps no vessel poses a greater potential risk for causing environmental damage than a laden tanker. However since the ***Exxon Valdez*** incident in 1989, significant changes have been made making tankers inherently safer, most notably the adoption of mandatory double-hull construction. Major advances in the field of tanker escort towing have created a new generation of purpose-designed, powerful tugs in this service. One can therefore state with confidence today that an escorted tanker, subject to a well-managed set of escort procedures, has an absolutely minimal risk of a cargo oil spill. Regardless of these safeguards, many political and environmental spokespersons cite the ***Exxon Valdez*** and similar incidents as reasons today for not permitting the passage of tankers through coastal waters. Much has been written about the resistance of double hull structures to collision, nevertheless there is a need to present this material in a way which is accessible to an interested public so as to help debate proceed on the basis of fact. This paper distils some of that data to a level which will contribute to informed debate. This includes an analysis of the speed and direction of impacts involving tankers which would be required to breach the inner cargo tanks. Work from Estonia, the United States, and Korea is examined demonstrating the worldwide interest in this topic. Using the example of the Port of Vancouver (Canada), the authors present the scenarios used locally for safe tanker operations, safely escorting tankers, and identify the potential risks and the associated potential for a cargo oil spill.

NOMENCLATURE

ASD	Azimuthing Stern Drive
AMSA	Australian Maritime Safety Authority
BP	Bollard Pull
BV	Bureau Veritas
CSR	Common Structural Rules
FPSO	Floating Storage, Production and Offloading
FSO	Floating Storage and Offloading
HPP	Harbour Practices and Procedures
HTS	Higher Tensile Steel
IACS	International Association of Classification Societies
km	kilometres
LNG	Liquefied Natural Gas
LNGC	Liquefied Natural Gas Carrier
LOA	Length Overall
MARPOL	International Convention for the Prevention of Pollution from Ships
MRA	Movement Restriction Area
OPA90	US Oil Pollution Prevention Act 1990
PSPC	Performance Standards for Protective Coating
VHF	Very High Frequency
VLCC	Very Large Crude Carrier

1. INTRODUCTION

Perhaps no vessel poses a greater potential risk for causing environmental damage than a laden tanker. However since the ***Exxon Valdez*** incident in 1989, dramatic and significant changes have been made to make tankers inherently safer, most notably the adoption of mandatory double-hull construction. The phase-out of single hull tankers now means that the vast majority of the world's oil-carrying fleet is double hull. Only some single hull storage solutions remain. Concurrently, major advances in the field of tanker escort towing have created a new generation of purpose-designed, powerful tugs for this important service. One can therefore state with confidence today that a tanker under the watch of properly designed escort tugs, and subject to a well-managed set of escort procedures has an absolutely minimal risk of a cargo oil spill. Regardless of these safeguards, many uninformed political and environmental spokespersons continue to cite the almost thirty year old ***Exxon Valdez*** and similar incidents as reasons today for not permitting the passage of tankers through coastal waters. Much has been written about the resistance of double hull structures to collision, nevertheless there is a need to present this material in a way which is accessible to an interested public so as to help informed debate proceed on the basis of fact and not hysteria. The use of escort tugs for safeguarding tankers is a topic not at all well understood by the general public and also not always properly understood by many within the marine industry itself. The advances made in this technology are dramatic in the years since the introduction of OPA'90, and tugs which incorporate the best available technology in this field have the capability to provide an incredibly high level of safety and operational redundancy to a transiting tanker, when properly deployed and when used with trained crews at all levels of involvement. This paper sets out with the objective of presenting the current state of the art in both these critical areas of ship design in a manner which can readily be understood by those interested in the facts and science of these important aspects of safe tanker operations.

2. RISK MITIGATION IN TANKER MOVEMENTS

2.1 RISK MITIGATION FACTORS

Oil spills from a tanker, though extremely rare, can occur in any of several possible ways; it could suffer an engine or steering system failure and then run aground and rupture the inner cargo tanks; it could suffer a structural failure in heavy seas; it could be involved in a collision with another ship resulting in ruptured cargo tanks, or human error could cause a navigational error resulting in a grounding. The probability of any such incidents is extremely low, but can never reach zero. More probably an error in a loading or cargo transfer operation could result in a spill at a terminal. An incident in the open sea cannot be protected against other than through diligence and vigilance in vessel maintenance and operations. However on the positive side, if a mechanical failure does occur deep sea then there is ample time to either execute repairs or to call for a capable towing vessel to assist. Time is on the right side of the equation in such cases. However once a tanker enters coastal waters time becomes the enemy, and any disabling event has the potential to result in a grounding and become a potential spill if suitable safeguards are not in place. The obvious safeguards to minimize grounding and associated pollution risks, in order of importance, are:

a) Safe Routing:

Allow tankers to operate only on those routes which provide a maximum sensible distance from shore, so that a known available response capability such as a rescue tug can respond to a breakdown and be on site well before grounding might occur. The "safe offset distance" for tanker routing can be readily calculated by conducting a relatively straightforward drift analysis which calculates the time it would take a drifting tanker under a defined set of environmental and tidal conditions to reach a set boundary line (NOT the shoreline, but say 1 to 2 km off the coast). That is then the maximum allowable response time for a rescue towing vessel to reach the ship and get connected to take the ship under tow. Such drift analyses also clearly help to define where rescue towing assets can be most strategically deployed.

b) Operating Procedures:

Responsible port management will put in place controls over all shipping within the area in which a tanker will operate. This should include ship traffic controls and priorities, restrictions on movement conditions, speed controls, use of pilot vessels, the use of qualified pilots, restrictions on time of movements where tidal influences may be strong, and the use of capable escort-rated tugboats.

c) Tug Escorts:

In areas where there is not sufficient sea room or response time to prevent a disabled ship from grounding under typical ambient conditions, a tethered escort tug must be used to continuously be ready to act as the backup steering and braking device for the tanker. In those locations where there is sufficient sea room the tug may not necessarily have to be tethered, but the time then taken for connecting results in far greater off-track ship movement. Careful analysis of all the potential failures and associated escort scenarios must be done to ensure the correct tug force generating capability is specified.

d) Escort Procedures:

Because disabling events aboard modern tankers are relatively rare, the use of an escort tug to steer and brake the ship must be practised routinely such that the reactions of the ship and tug crews are instinctive if and when a real incident does occur. Tug Masters must understand in which instances it might be better to assist a turn rather than oppose it. They must also be acutely aware of the limits of safe operation for the tug itself.

e) Spill Response Capability:

As unfortunately no set of precautions can assure 100% risk reduction, one must ultimately be prepared to respond to a spill incident. This is more likely to occur due to human error than to a mechanical/structural breakdown of a tanker. In either case however it requires providing the equipment, response organizations and associated protocols, and the trained personnel to operate the equipment in the worst foreseeable conditions. Assets must be deployed in positions of maximum strategic benefit.

2.2 TANKER CONSTRUCTION

The physical cross section of a modern double hull oil tanker is largely defined by the requirements of both OPA 90 [1] and MARPOL [2]. OPA 90 sprang from the grounding of the ***Exxon Valdez***, a single hull VLCC, in Prince William Sound, Alaska in March 1989. The consequences of this accident, aside from the pollution, were long lasting changes to the design of oil tankers, principally increased protection of the cargo space by the introduction of a void space surrounding the cargo area. When the ship is light, in ballast without cargo, this space is used for ballast water so as to maintain a safe draft and allow the ship to operate efficiently.

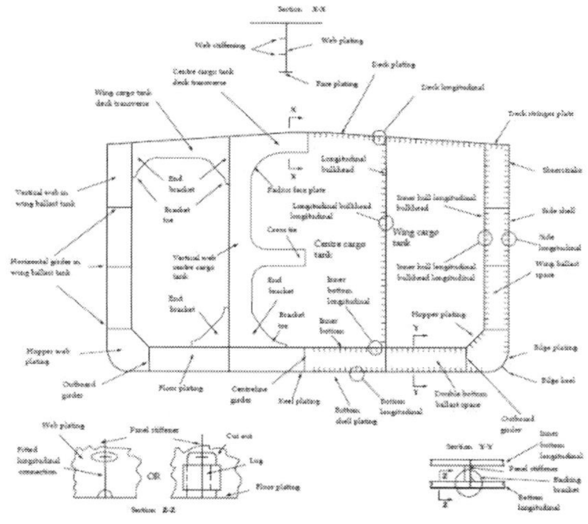

Figure 1: Typical double hull oil tanker midship section (source IACS)

Figure 2: Double hull tanker under construction

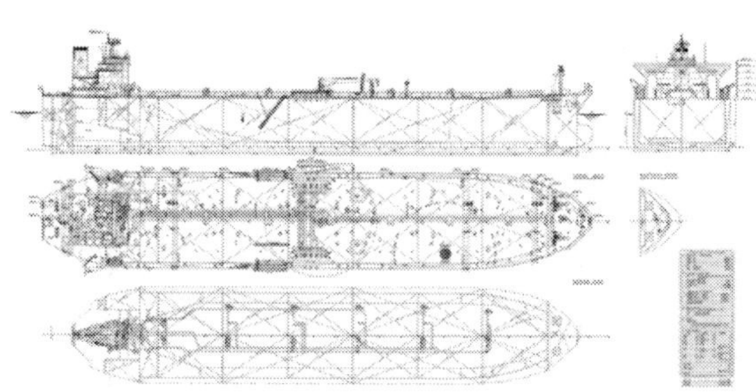

Figure 3: Typical general arrangement for a double hull tanker

Figure 4: Typical view over fore deck of a double hull tanker, line of double side is about 2.5 m in from ships' sides, beam about 48 m

Figure 5: Typical oil tanker double bottom internals, about 2.5 m height

The effectiveness of double hull tankers in reducing the risk of pollution was heavily debated during the development of both OPA 90 and the MARPOL double hull amendments. The basic rationale used to support double hull requirements at both the federal and international level is that double hulls reduce the risk of oil spills that occur during a low energy grounding or collision. Because most accidents occurring in or near ports typically involve lower vessel speeds, and because the risk of grounding or collision is typically highest in port areas, double hull tanker designs offer a very reasonable option for reducing oil spill risks under these circumstances ([3] Elise DeCola, 2010).

A typical oil tanker of the type that might be seen to be loading at the Kinder Morgan terminal in Vancouver will have a double hull width of at least 2 m; that is 2 m at the side and 2 m at the bottom. The minimum, regardless of ship size, is 1 m for double hull and double bottom protection of cargo tanks. The minimum MARPOL requirement for double hull side width and bottom height is almost invariably exceeded in practise because of the need to carry adequate ballast in the double hull spaces.

2.3 TUG ESCORTS IN CONFINED WATERWAYS

The legislated use of tugboats as tanker escorts really began with the introduction of OPA 90 in the USA. Shortly thereafter many jurisdictions enacted similar legislation. However at that time the tug world really did not have a full appreciation of what was involved in tasking tugs to provide steering and braking capability of the magnitudes required to control large tankers at speeds of up to 10 knots. The topic was discussed at length at many International Tug Conferences. There was no shortage of those who stated that this simply could not be done. An example of just such a claim was from Ken Ross [4], at the time a senior executive in a major international towing company who stated emphatically; *"... the real crux of the problem ...is that in reality undue reliance is being placed on the escort tug to provide a cure-all.... such reliance is absolute pie-in-the-sky and will depend more on luck than good judgment and management."* He went on to say; *"...there is virtually unanimous opinion that the tug which can always be relied upon to bale (sic) a runaway or disabled ship out of trouble does not exist."*

Captain Ross' comments might still be accurate if the state of tug design had not evolved rapidly to accept head-on the challenge of safe tanker escort. Extensive private testing and research into better hull forms for escorting, and the associated generation of higher escort steering and braking forces while still ensuring tug safety certainly did take place over the ensuing years. Much of that work has been well documented in the proceedings of the International Tug & Salvage conferences, including extensive work by the Author's company. Important references in this context also include papers by Banks [5], Laible & Gray [6], Amundsen and Rokstad [7], and Allan and Molyneux [8].

The result of this work is a new generation of highly capable and efficient escort tugs, designed exclusively for this challenging work, and able to work safely in the "indirect mode" under high lateral forces without fear of capsize or girting. The use of sponsoned hulls such as the ***RAstar*** form developed by Robert Allan Ltd. in the mid-90's (Figure 6) enables these tugs to operate at heel angles of 12° to 15° with higher initial stability than they possess in the normal upright condition.

Figure 6: ***RAstar*** sponsoned hull form initiated with VSP escort tug – ***Ajax***

The use of properly-designed escort tugs in restricted waters thus offers the capability of providing emergency steering and braking capacity equal to that of the ship itself, in effect bringing a full degree of redundancy to the ship propulsion and steering systems in the event of a breakdown in critical areas. Ideally computer simulations should be conducted which define the vector forces to be applied by an escort tug in order to affect a "save" of a disabled ship within pre-determined constraints in the channels to be navigated. These forces may be greater than those which the ship itself could generate, particularly in very narrow channels.

2.4 CONTROLLED SHIP MOVEMENTS IN RESTRICTED AREAS

Most jurisdictions today have in place strict protocols around the movement of tankers, or indeed for any vessels carrying dangerous goods. The controls required vary naturally according to the individual port or terminal, but will generally include the following:

- Limitations on other vessel movements during tanker moves
- Speed controls on all ship movements
- Use of "pilot" vessels ahead of tankers to warn other vessel operators (especially small craft operators)
- Use of one or more qualified Pilots aboard the tanker
- Definition of shipping lanes for inbound and outbound traffic
- Defined use of properly rated escort tugs, with specified tug performance and tug deployment configurations
- Agreed communication protocols

2.5 TRAINED CREWS

All the best equipment in the world will neither prevent an accident nor facilitate a recovery if the persons involved in all aspects of the operation are not fully trained to deal with all foreseeable scenarios. Tanker rescue towing and indeed oil spill recovery are "firehall" scenarios; one hopes that these services will never be required but as long as there is some small potential for an incident the crews aboard these vessels must be fully trained to respond. Tanker escorts are in a different category: these are activities that take place regularly, thus providing the opportunity for the tanker crew, pilots and tug operators to conduct simulations of steering or propulsion failures and learn the best tactics to follow in response to each. They also learn the inter-dependencies of the actions by all these parties. These exercises make all involved aware of the expected response of the ship to the tug actions, and thus they know what can be expected should a real incident occur. Tug crews in particular must be ready to act at a moment's notice, and the response actions must be second nature to them. Such familiarity is only possible with: (a) proper training to understand the limits of the capability of the tug (b) training in the range of escort response strategies available, and then (c) conducting regular exercises and establishing clear communication protocols for all the potential scenarios to identify any sensitive areas of operation.

3. DOUBLE HULL STRUCTURES

Work looking at the collision resistance of ship structures is generally recognised as having been started by Minorsky [9] in 1959. This early work of course took no account of double hull tanker structures as these remained still to be invented. More recent studies related to the design technology of ship structures against collision and grounding have been reviewed by J.K. Paik [10] who has worked extensively in the field of damage resistance of ship structures. A brief review of key work in the field follows, there is certainly a lot that has been done.

Arita (1986) [11] carried out theoretical, numerical and experimental studies on the structural crashworthiness of nuclear powered vessels, LNG carriers and double hull tankers in collision. A theoretical approach to predict the energy absorption capability of the colliding and collided vessel structures was proposed and explored.

Kierkegaard (1993) [12] developed an analytical method for analysis of the internal collision mechanics of the bow structure in a head-on collision with icebergs or other objects.

Wang (1995) [13] obtained a simplified theoretical approach for predicting the damage of vessels in a head-on collision or in raking. A variety of failure modes were considered in Wang's work.

Lee (1997) [14] developed a theoretical formulation for estimating the damage of tanker structures in grounding with a forward speed or in stranding at standstill. The Lee method is formulated in closed form, and takes into account the vertical crushing caused by ship rigid body motions as well as horizontal raking (i.e. cutting) caused by the forward movement of vessel.

Simonsen (1997) [15] developed a mathematical model for predicting the loads and hull girder response of vessels in grounding on a soft sea bed or on a rock pinnacle. Simonsen's contribution is of value to the structural design of fast vessels against collision and grounding.

Chung and Paik (1997) [16] developed a theoretical approach for analyzing the damage and crushing strength of a ship's bow in a head-on collision. A series of crushing tests under quasi-static and dynamic loading were carried out on thin-walled square tubes with axial and/or circumferential stiffeners, and also on unstiffened specimen. More recently, Zhang (1999) [17] presented a comprehensive set of simplified formulations for the analysis of structural damage in ship collisions and grounding.

Further work was developed by Alan Brown [18] et al. examining structural response to collision and grounding and is particularly important to the current debate in western Canada because the size of ships examined is broadly in line with the types of ships that operate in the Salish Sea and Vancouver Harbour. Additional work carried out by Pollution Prevention & Control (POP&C) at Herbert Engineering in 2004 to 2006 [19] focused on Aframax tonnage and this is absolutely directly applicable to Salish Sea energy transportation. Work in the field has been brought up to date by Heinvee and Tabri [20] in Estonia in 2017. In addition Paik [21] has continued to work in this area and has published extensively including structural crash worthiness.

At its simplest the behaviour of ships' structures in a collision, a ship on ship impact, is best described in energy terms. Most of the discussion here will consider a bow from a striking vessel hitting the side of a struck vessel. Other collision types are possible, such as bow to bow (head on) or bow to stern in a failed overtaking manoeuvre or bow to accommodation (see Figure 7).

Figure 7: Accommodation damage from bow impact (RNLI)

However these scenarios are much less likely to result in penetration of the cargo space with consequent loss of oil than a bow to side collision. When the bow of a colliding vessel strikes the side structure of a collided vessel, the initial kinetic energy may be consumed in many ways. Part of the kinetic energy will be dissipated by damage of the colliding and the collided vessel structures, while the rest of the kinetic energy will remain in the form of various movements of the vessels together with the surrounding water.

4. PENETRATION LEVELS

Various technical efforts have been made to establish penetration distances after collision and grounding and some of those will be discussed here.

Sandia Report (Ammerman, 2002) [22], Sandia had previously conducted sophisticated finite element modeling of collisions of a series of ships with a double-hulled oil tanker similar in overall size, mass, and design to an LNG vessel. A summary of the analysis of a 90° collision of a large container ship (50,000 metric tonne class ship) and a double-hull tanker (80,000 metric tonne class) is shown in Figure 8 and collisions with smaller ships are shown in Figure 9 (Ammerman, 2002). The analysis tool included an approximately 250,000 finite element model of both the impacting vessel and the double-hulled tanker using PRONTO-3D run on a massively parallel computer with 256 processors. This is a transient dynamic, explicitly integrated, Lagrangian solver of the equations of motion. The analysis tracked the progressive failure of the struck ship as the striking ship penetrated. As noted in these figures, breaching of the inner hull does not occur until impact velocities exceed approximately 5 to 6 knots for large vessels. For small vessels, such as pleasure craft, kinetic energy is approximately one to two million N·m. Figure 9 shows that this level of kinetic energy is generally insufficient to penetrate the inner hull of a double-hulled vessel such as an LNG ship.

This analysis also calculated that penetration into a double-hulled tanker must be approximately three metres before a hole occurs in the inner hull. This, therefore, can be used to estimate the minimum size of an opening in the outer hull to likely cause a penetration in the inner hull and a spill in a grounding or collision event. The results for this analysis were compared with initial collision information from the ***Baltic Carrier*** collision in 2001 at approximately 12 knots. The results of these analyses over-predict, by about 15%, the external hole size measured for that collision. ***Baltic Carrier*** had adeadweight of about 37,000 tonnes, significantly smaller than an Aframax tanker and with a correspondingly reduced side protection.

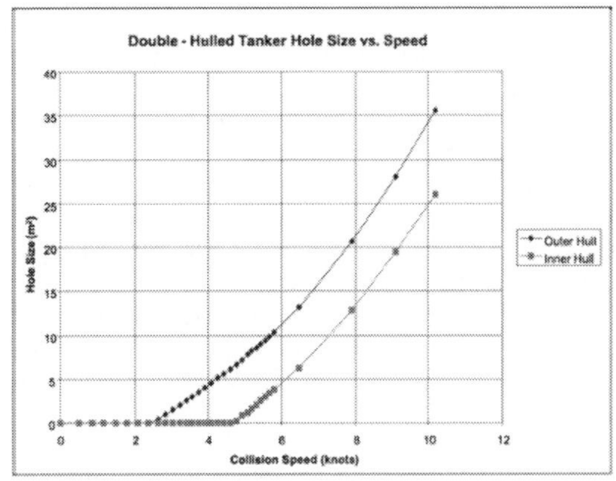

Figure 8: Study estimate of speed required to create a given hole size

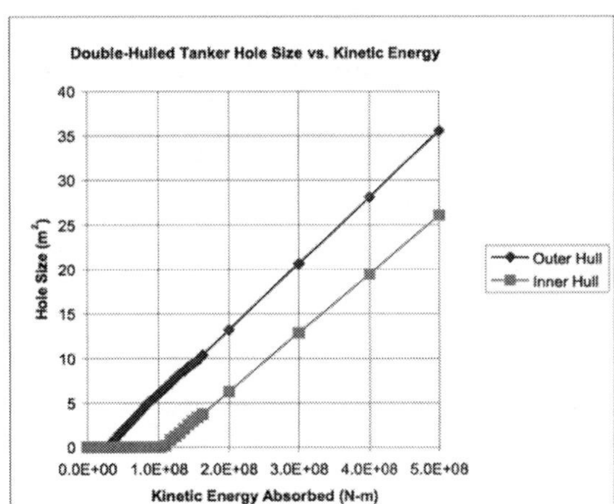

Figure 9: Double-hull tanker study of energy required to create a given hole size

Figure 10: ***Baltic Carrier*** after collision, 2001 (photo ITOPF)

An additional element to consider in the accident scenario is that the hole size developed probably is not the size of the spill orifice. In many collisions between two ships, the ships can remain joined for several hours if significant penetration of one ship occurs. The analysis by Ammerman suggests that as little as 5 to 10% of the generated breach size is available for leakage. Therefore, the collision of a large ship with an oil tanker at even 12 knots is expected to produce an effective hole area of no more than approximately one square metre for an oil spill. If larger spills do occur, hole sizes could approach those calculated for intentional breaches. Grounding and terrorist attacks, for example, could produce holes that are effectively larger than in collision as the there is no ship to plug the opening in the initial stages. Analysis suggests that, in most cases, an intentional breaching scenario would not cause, in the case of LNG, an LNG tank breach of more than 5 to 7 m^2.

Table 1: Estimated LNG ship damage from potential tank breaches and spills

Breach Event	**Breach Size**	**Tanks Breached**	**Ship Damage**
Accidental collision with small vessel	None	None	Minor[b]
Accidental collision with large vessel	5-12 m^2 (Spill area 0.5-1 m^2)[a]	1	Moderate[c]
Accidental grounding	None	None	Minor
Intentional breach	0.5 m^2	1	Minor
Intentional breach	2 m^2	1	Moderate
Intentional breach	2 m^2	3	Moderate
Intentional breach	12 m^2	1	Severe[d]
Intentional breach	5 m^2	2	Severe

Notes:

a: Assumes vessels remain joined during spill event and breach is mostly plugged.
b: "Minor" suggests ship can be moved and unloaded safely.
c: "Moderate" suggests damage that might impact vessel and cargo integrity.
d: "Severe" suggests significant structural damage. Ship might not be able to be moved without significant difficulty and includes potential for cascading damage to other tanks. Cascading events are unlikely in oil spills, these can occur in LNGC because of the cryogenic temperatures involved and the effect on steel.

In LNGC the potential exists for progressive structural damage due to cold shock from the cryogenic liquid on the structural steel of the ship. The extent of the damage will depend on the volume and rate of LNG spilled and the ship areas that will be directly contacted by the liquid LNG. Such damage happens very quickly. Oil tankers are far less prone to cascading damage than LNGC since there is no mechanism for imparting sudden thermal shock to the structure. Based on the postulated breach events, attempts were made to estimate the potential level for ship damage from both accidental and intentional events. These are presented in Table 1 above. The conclusion from this table has to be that mitigating measures are required including tugs, VTS, collision avoidance and navigation since a double hull alone is not enough to prevent tank breaches, rather the double hull and escort towing and the other mitigation measures must work together to provide a virtuous circle of protection.

Heinvee and Tabri (2017), more recent work has been carried out by Heinvee and Tabri [20].This is interesting as it tries to take limited parameters such as length and displacement and then develop criteria to determine whether or not the inner hull of a double hull tanker ruptures in a collision. Part of the interest in the paper is to examine the impact of higher tensile steel on collision resistance. Contemporary ship designs are seeing increasing use of higher tensile steels. Increasing yield strength gives higher deformation energy however the consequence of using HTS is that plate is thinner, naturally. On the whole the yield strength does not have an impact on collision resistance, the effect is largely neutral. In principle this allows for ready comparison to be made between older pre CSR DH tankers and more modern tankers where HTS is being used in a much higher percentage than in the older ships.

Table 2: Results (striking ship velocity v_A 2.5 m/s, 4.9 knots) comparing FEM analysis with Heinvee and Tabri's new approach

Scenario								Coef.of energy absorbtion.	FEM		New approach		
Ship B ←V_A Ship A	L_B	L_A	T_B	T_A	h_{dw}	Δ_B	Δ_A		E_{def}	δ/h_{dw}	E_{def}	δ/h_{dw}	P
	[m]	[m]	[m]	[m]	[m]	[ton]	[ton]	[–]	[MJ]	[–]	[MJ]	[–]	[-]
(1)150–150	150	150	8	9.5	1.2	20 910	24 867	0.53	53	1.75	43	1.48	0.69
(2)150–150	150	150	8	5.7	1.2	20 910	14 898	0.65	20	1.17	32	1.95	1.00
(3)150–190	150	190	8	11.9	1.2	20 910	55 009	0.34	33	1.58	61	2.17	1.00
(4)150–190	150	190	8	14.2	1.2	20 910	65 730	0.30	56	1.83	64	1.78	0.97
(5)190–150	190	150	12	10.4	1.8	55 621	27 186	0.73	69	1.11	65	1.21	0.23
(6)190–150	190	150	12	6.9	1.8	55 621	17 907	0.81	57	1.33	47	1.18	0.18
(7)260–150	260	150	18	6.9	1.8	130 478	18 037	0.91	64	1.17	54	1.06	0.07
(8)260–150	260	150	18	12.5	1.8	130 478	32 703	0.84	84	1.33	90	1.57	0.81
(9)260–260	260	260	18	18.8	1.8	130 478	136 140	0.56	142	1.33	251	1.87	0.99
(10)260–260	260	260	18	14.0	1.8	130 478	101 418	0.63	86	1.39	210	2.48	1.00

Striking velocity is taken as 10 m/s in the numerical simulations which is about 19.5 knots. This is conservatively high for the confined waters of harbour approaches and some pilotage areas where speed controls can be expected. Salish Sea pilotage areas are considering reductions to 17 knots for large container ships and further reductions in congested areas according to the December 2016 Salish Sea Oil Spill Risk Mitigation Workshop held in Washington State.

In the four step process developed in Heinvee and Tabri's paper a model speed of 2.5 m/s is used, (Table 2), or about 5 knots which is below the maximum speed for pilot boarding in Puget Sound, Washington State according to the Puget Sound Harbor Safety Committee. The model can however take any speed and this would be useful for further investigation. The model would benefit from some more contemporary double hull models, for example there is no lower sloping hopper plate in the struck/striking ship model, the double hull itself is less than 2 m wide and the hull itself is rather slender; breadth to depth is 1.4 whereas more modern designs with a similar length to the T260 in the first column in Table 2 have B:D of about 2.2 with a double hull width of 2.7 m.

Some explanation of Table 2 will be helpful: ∂_f/h_w is penetration depth over double hull width; where this is greater than 1, penetration of the inner hull has taken place. Given the constraints imposed by the model type, slenderness and narrowness of double hull, then it is to be expected that running the model with more representative figures will show that the double hull can resist low speed collisions where the striking ship is up to 260 m long and striking at right angles. Other angles, relevant for passing ships should be examined.

Traditionally modelling has included a bulbous bow on the striking ship. Contemporary designs of bulk carriers, tankers and to a lesser extent LNGC feature a much reduced bulb and this is likely beneficial for impact.
POP&C (Pollution Prevention and Control) (Herbert Software Solutions, 2006) [19], this European Commission study which ran from 2004 to 2006 looked at survivability performance of medium-sized oil tankers and at the time of writing needed to be based on historical data dominated by single hull vessels. The database consisted of events such as collisions, contacts, groundings, fires and explosions. The study does show that medium size tankers have a high rate of survivability following loss of watertight integrity and this is known to be true intuitively. Where a fire occurs then there is a high probability of serious consequences and it can be seen that this is supported by the recent collision off Shanghai of the tanker ***Sanchi*** and bulk carrier ***CF Crystal***.

Of more interest is the commentary around outflow following collision. Among other things the report notes that Aframax tankers have a high rate of survivability following loss of watertight integrity and this accords with experience. The investigation performed case studies and compared historical data with analysis. Given that the historical database for this analysis stopped in 2003 it is high time that this study was updated. Interestingly though for all collisions/allisions examined there is a 0.68 probability of zero outflow, or conversely a 0.32 probability of some outflow. It can also be seen from the statistics presented in the report that double side or double hull significantly reduces the outflow/capacity ratio in any incident.

Some discussion of oil outflow is appropriate here where damage spans the waterline. In this case all oil in the damaged tank, provided it is lighter than water will eventually be lost. For the case where damage is all above the waterline then oil will flow out to the bottom of the damage. With damage entirely below the waterline this can be modelled simply, as outflow will continue until hydrostatic balance is reached.

5. ALLISION AND GROUNDING

Figure 11: Sideshell damage to moored tanker (allision), note inner hull remains intact

While most of the focus is on collision there are also incidences of allision where, for example, moored ships can be contacted by passing traffic or a ship can contact a bridge. There is a risk of a spill in both allision and grounding and these incidents should not be overlooked. For the case of the moored ship being the struck ship there is probably less likelihood of a spill than for an open sea collision. First of all, passing speeds are likely to be low and fully laden ships do not remain alongside so any accident is likely to involve a partly laden tanker.
The POP&C report [19] also discusses grounding and looks at the probability of zero outflow for a grounding incident. Here the probability of zero outflow is 0.42 however it also has to be said that all the ships that grounded and had greater than zero outflow were single hull. In bottom damage oil is constrained from flowing out by seawater hydrostatic pressure. There is also an element of dynamics at play in grounding and a high speed grounding can be expected to release oil immediately if the double hull is penetrated. Modern tanker designs have all oil tanks protected by void spaces, including bunker and service tanks in the machinery space, an additional environmental enhancement that was not fully considered by the authors of the POP&C report.

The report also discusses the development of a survival probability index, the greater the number the more survivable is the particular tanker design. Tankers with double bottom only fare particularly badly confirming the need for a full double hull where grounding and collision are possibilities. Conversely FSO/FPSOs could be built with a double side only where there is no possibility of a grounding while loaded.
Reference [18], Brown et al's 2000 work, notes that the further from the centreline of the ship that a rock is contacted the greater the damage, a function of relative motion. This can be directly translated into ensuring that navigable channels have adequate breadth and that survey lines are tight to avoid missing erratic rocks.

In the same work it was demonstrated that tank spacing and bulkhead characteristics have little effect on damage results whereas thickness of outer bottom plate did have an impact. Contemporary rule-based construction for oil tankers adds a corrosion margin to the design and together with PSPC requirements it can be expected that later double hull designs have an increased resistance to grounding damage simply by design, an unintended, though welcome side effect.

Further investigation of midship section enhancements showed that fitting two additional girders in the double bottom did have a marked effect on grounding resistance. This though comes at a steelweight penalty and a recent survey of a 120 ship tanker sample showed only four ships to have been fitted with additional girders.

6. TYPES AND CAPABILITIES OF MODERN ESCORT TUGS

In the early 1990's, as the concept of a true escort tug was just beginning to evolve, the use of Voith Schneider cycloidal propulsion was considered to be the "Best Available Technology" (B.A.T.) for escort tugs, and in many respects was considered to be almost the only available choice of propulsion. Owners however were seeking more flexibility in these large and relatively expensive tugs, and the use of Z-drive propellers in the "azimuthing stern-drive" (ASD) configuration was considered highly desirable in not only offering a higher unit thrust/kW but representing a configuration of tug which was more suited to other tasks such as long-distance towing or even anchor-handling in the event that an escort contract was terminated. Unfortunately the earliest generation of ASD escort tugs were not at all well-suited to escort service. Figure 12 from Aarts [23] shows an example of a typical ASD terminal tug of this era. While these were considered very good general towing and ship-assist tugs, they were not well-suited to the role of tanker escort. In fairness, these were not conceived to be so, but tugs of the same general configuration, only larger, were used in the early to mid-90's for tanker escort. They lacked any high-lift skeg to generate hydrodynamic forces; their aft-biased skeg, fitted for general course stability, was a significant hindrance to effective escort and the towing point was much higher than desirable for this service. Many of these "conventional" tugs also had stability characteristics much less and freeboard lower than necessary for safe escorting. The end result was that ASD tugs were by the late-90's deemed by many in the industry to be unsuitable for use as escort tugs. The rationale was incorrect; it was not the type of propulsion which was at fault but simply the wrong hull design being applied to the task.

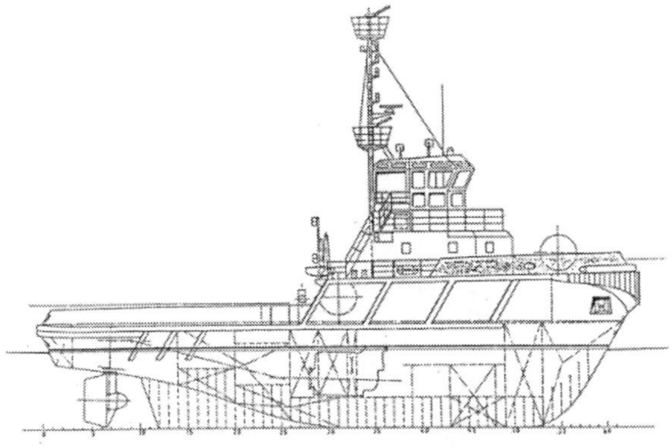

Figure 12: Typical ASD Terminal Tug circa 1990

Extensive research throughout the 90's by Robert Allan Ltd. gave rise to the use of much superior, sponsoned hull forms for escort duty, for tugs with both VSP and Z-drive propulsion systems, namely the aforementioned ***RAstar*** hull form. The incorporation of large, high-lift, forward-biased skegs (previously unknown on ASD tugs) significantly increased the indirect force generating capability. Intelligent design increased freeboard and stability throughout the operating range for a new generation of ASD escort tugs which began to emerge in the late 90's, as illustrated in Figures 13 through 16. Most critically however, the early stigma attached to ASD escort tugs was erased and the latter have since become the norm in this field, with widespread acceptance internationally.

Figure 13: ***Z-Tech*** Class tug ***Indee*** indirect at 9 knots

Figure 14: ***RAstar 3400*** Class tug ***Svitzer Haven*** in heavy weather

Figure 15: ***RAstar 3900*** Class tug ***Svitzer Kilroom***

Figure 16: ***RAstar 4000DF*** Class dual fuel terminal escort tugs

Other hull form/drive configurations have also emerged latterly to fill needs in this market. The triple Z-drive Rotortug® has emerged as a powerful and effective option for tanker escort. After extensive evaluation and comparative simulations in studies for the proposed Northern Gateway Pipeline project in northern British Columbia a Rotortug® configuration was chosen as the best option for the route and conditions. The resulting design concept for the "***RAincoast*** Guardian" Class tugs (Figure 17) would have been the largest and most powerful escort tugs afloat anywhere. Political and environmental opposition however caused that project to be abandoned. The technical work which went into that design however has not been lost and is being applied to similar projects.

Figure 17: ***RAincoast Guardian*** Class escort tug concept for BC North Coast

Another significant development is the ***RAVE*** design concept for escort tugs. Based on observations made during various VSP tug tests, it was anticipated that using two VSP drives in an axial rather than in the typical side-by-side "tractor" configuration would provide better efficiency and higher escort performance. Collaboration ensued between Robert Allan Ltd. and Voith GmbH & Co. KGaA, including extensive CFD, simulation and model testing work to prove this unique concept. Although it is a very specialized configuration (Figure 18) the test results indicated very superior performance. The concept has also been adapted to more conventional ship-handling tug operations, and in early 2018 the first ***RAVE*** tug was delivered (Figure 19), also incorporating the patented Carrousel towline handling system [24]. While the latter mechanism is presently not sufficiently robust for large ship for escort service, the concept is certainly adaptable.

Figure 18: ***RAVE 3700*** Class escort tug

Figure 19: Carrousel ***RAVE*** tug on trials

As these new and much improved design concepts began to emerge, operators also began to examine and experiment with the most effective modes of tug operations for escort manoeuvres. The "science" of escort towing was gradually emerging and was being better understood. The forces involved in escorting are extremely speed dependent, and thus the mode of operation varies as the tanker speed increases. Obviously at zero speed a tug can only apply direct bollard pull, and up to about 5 knots a direct pulling mode produces the most net line pull. In the intermediate speed range however, say from 5 to 7 knots, ASD tugs can oppose their drive units and generate forces higher than direct thrust in the so-called "transverse arrest" mode. Finally, as speed increases to 8 knots and above tugs move into the "indirect" mode and use the hydrodynamic forces generated by the hull and appendages to produce forces which can frequently reach more than 1.50 x the bollard pull. The ability to achieve such high levels of line force is very dependent upon the stability characteristics of the tug however, and not all tugs are suitably designed to reach this level of performance.

Coupled with these design advances, there have been significant improvements and most thankfully some harmonization among Class Societies in the regulations associated with the stability of escort tugs, as attested to by de Jong [25]. The current basic regulatory requirements formally introduced in 2017 include the following:

- Establishing of service notations which distinguish the type and location of tug operations, with differing criteria for each
- Formalization of escort towing stability criteria based on righting energy, and accounting for all forces acting on the tug
- Accounting for the influence of the longitudinal separation of tow point and thrust point
- Acceptance for certification of valid escort simulations rather than a full-scale escort performance trial

- Formalization of the requirements for towing gear and the relative strength of each component of the towing system, and finally
- Specific mention of the requirements for adequate towing fittings on the attended ship

Of critical importance in the development of safer and more capable escort tugs is the development of hawser winches able to render and recover towline under high load, avoiding wave or manoeuvre-induced peak loads which might otherwise result in broken towlines. This reflects a significant change in philosophy for towline strength; in previous generations of tugs it was assumed that in the event of a tow getting into trouble, or indeed the tow putting the tug into trouble, the crew were given an axe with which to sever the towline, leaving the tow to its own fate! The futility of this regulated requirement can be attested to by many a tug girting in prior years, events which rarely if ever occurred in a controlled manner which would enable crew to deliver the required axe blows. Today's escort winches provide the ability to sustain defined line tension throughout a wide range of operating conditions at values which will not put the tug at risk of capsize, and which equally will not sever the connection between the tug and the attended tanker. If overloaded the winch simply lets line out at a controlled rate and then recovers to the desired length when tension is reduced. Of course the associated advances in high-strength towline technology have accompanied the improvements in winch design, and designers should be paying attention to the tug fittings required to take maximum advantage of new rope technology, as well as methods for coping with the high heat which can be generated in these lines under tension. Figure 20 illustrates a cooling water spray system installed on the dual aperture staple of a modern escort tug. The "wide-A" format of this staple (lower aperture) also reduces the heeling moment under load.

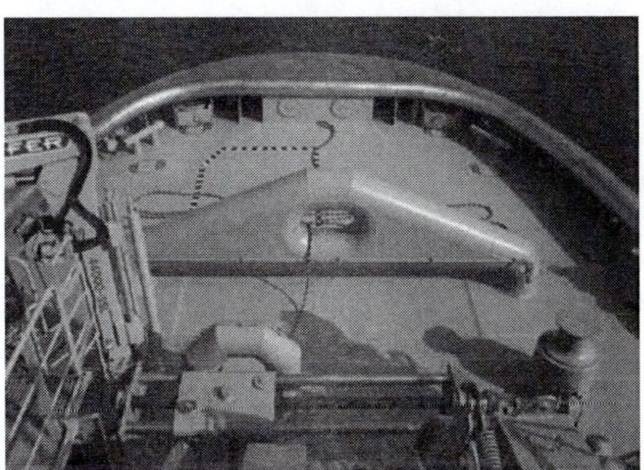

Figure 20: Escort towing staple on *TRAktor-V 3600* Class tugs

In summary, it is more than fair to state that the design of modern tanker escort tugs have evolved rapidly and significantly over the past 20 to 25 years to the point that the powerful machines available today represent the single greatest safeguard available to prevent a tanker mishap in confined waterways, when used within a properly planned and tested tanker escort plan. Escort tugs should be selected which can provide at the very least lateral and braking forces equal to those which can be generated by the attended ship's rudder and by the propeller respectively, or if available, equivalent to the forces shown by simulation/analysis as necessary to affect a "save" of a disabled ship within the confines of the channels to be navigated.

7. A WORKING EXAMPLE

There exist around the world today many examples of tanker escort systems which match the requirements stated above. Unfortunately there are also many which do not! One of the best examples of a fully integrated escort system is one of the very earliest, namely the Alyeska tanker escort systems in place both in Prince William Sound, Alaska and in Puget Sound at the southern terminus of that shipping system in Washington State. As a closed system the tankers are dedicated to that route and hence tug masters, ship masters, pilots and all other interested parties participate in regular reviews and tests of the system.

In Canada the scene is quite different and several proposed major pipeline/terminal projects receive much misguided resistance. The Port of Vancouver, Canada's largest port, currently only exports a modest amount of oil, with about 40 to 50 tankers per year transiting the harbour and associated sensitive waterways until they reach the open Pacific, a total distance of about 140 n. miles. A major oil pipeline project currently approved for construction and terminating within the Port boundaries will increase that volume of traffic to about 400 tankers per year...more than one per day on average. This increase has many local and Provincial government officials and environmentalists alike ranting about the risk of oil spills within the harbour as well as on any portion of the British Columbia coastline. Unfortunately these opponents often cite incidents such as the ***Exxon Valdez*** spill as examples of why such volumes of shipping or indeed any oil shipping should occur along this beautiful and sensitive coastline. While it is not unreasonable to insist that the very highest standard of precaution must be taken, it is too often the case that the need for "world class oil recovery systems" are cited by politicians and bureaucrats alike as the solution rather than placing an emphasis on preventing a spill in the first place. The facts of double-hulled tankers and the importance of escort tugs in this preventative role are rarely mentioned and certainly are much less understood. A review of the proposed operational guidelines and constraints follows with an assessment of its compliance with best practises.

Figure 21 illustrates the proposed voyage from tidewater to the open Pacific and the portions on which escort procedures will be in place.

Figure 21: Tanker escort route from Vancouver Harbour to open Pacific (source: https://www.transmountain.com)

The current "system" for safeguarding the Port of Vancouver waterways from potential pollution is defined within a document developed by the Vancouver Fraser Port Authority entitled "Harbour Practices and Procedures" (HPP). Section 8.13 of the HPP is the "Second Narrows Movement Restriction Area Procedures (MRA-2)", which includes the following most critical requirements for tanker movements within the "MRA" – (Movement Restriction Area). Figure 22 illustrates the MRA zone.

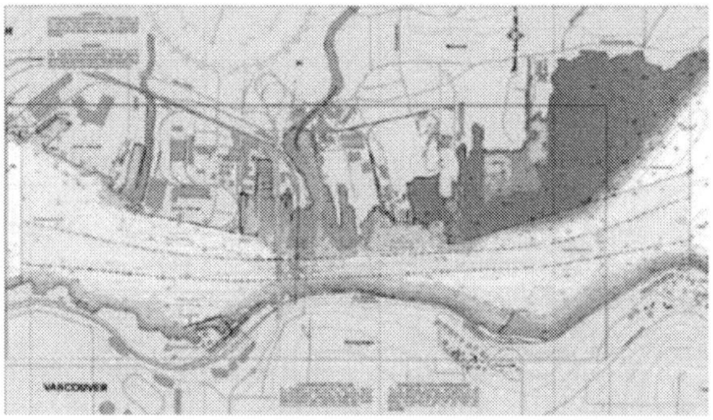

Figure 22: MRA-2 Zone in Vancouver Harbour

a) Transit Windows:

An MRA-2 Transit: is defined as a movement within MRA-2 that includes passing under the Second Narrows Iron Workers Memorial Bridge and the Second Narrows Railway Bridge. Transit windows are established on either side of high and low water slack tides and are based on predicted slack water or stemming a predicted limiting current of one or two knots.

b) Transit Restrictions:

The following specific transit restrictions and requirements apply:

- Tankers greater than LOA 185 metres and/or 40,000 Summer Deadweight (SDWT) and above are restricted to daylight transit of MRA-2 when in product
- Loaded tankers shall be trimmed to an even keel or by the stern and shall not be trimmed by the head.

c) Clear Narrows Restrictions:

The term "Clear Narrows" is defined as the transit of a vessel through MRA-2 unimpeded by any other vessel. MCTS will declare a "Clear Narrows" on VHF Channels 12 and 16 by means of a Sécurité call to ensure unimpeded transit of restricted vessels, examples being but not limited to:

- A vessel with LOA 230 metres and above and/or a moulded breadth of 35 metres or above
- A Tier 1 vessel (tanker) in product

d) Speed Restrictions:

Tier 1 vessels shall transit or manoeuvre within MRA-2 at a safe speed not to exceed six knots through the water, except when safety of navigation requires otherwise.

MRA-2 PILOTAGE REQUIREMENTS

Pilotage requirements within VFPA jurisdiction are governed by the Pacific Pilotage Authority Regulations, Section 9 and 10:

- Tankers of 40,000 summer deadweight tonnage (SDWT) and above in product require two pilots for an MRA-2 transit. Both pilots shall remain on the bridge throughout the transit. Two new pilots will replace the two shifting pilots in English Bay or other agreed location
- MRA-2 vessels with LOA 230 metres and above and/or a moulded breadth of 35 metres and above require two pilots for an MRA-2 transit

MRA-2 VESSEL ASSIST TUG REQUIREMENTS

MRA-2 vessels when transiting MRA-2, must comply with the standards for tug requirements included in Table 3 or 4 as appropriate, which detail the number of tugs and bollard pull requirements, reasonably spread between tug hulls. In addition:

- All vessel assist tugs employed on piloted MRA-2 vessels transiting MRA-2 must be tethered tractor/ASD tugs
- Vessel assist tugs capable of generating more than 40 tonnes of bollard pull shall have an operational tension meter that the tug operator can easily read from the conning position
- All vessels which require tethered tugs for an MRA-2 transit shall have them tethered

prior to entering MRA-2 and shall remain tethered until clear of the Second Narrows Bridges unless, for operational reasons, they are required to remain tethered beyond MRA-2

- Tankers of LOA of 185 metres and above in product and/or 40,000 tonnes SDWT and above in product require a minimum of two tugs that, when inbound must be tethered prior to transiting MRA-1 and when outbound shall remain tethered until clear (west) of MRA-1

An interrupted passage between Second Narrows and First Narrows bridges, for whatever reason, shall not reduce the minimum escort tug requirements for the transit. For escort and tethered tug requirements related to tankers of 40,000 SDWT and above in product outside of VFPA jurisdiction, reference shall also be made to the relevant Pacific Pilotage Authority Notices to Industry.

Table 3: Second Narrows MRA (MRA-2) transit procedures deep sea vessels – summary matrix

Tankers in Product

Vessel Type	Night Time Allowed	Tidal Current Opposing	Tidal Current Following	Tugs	Pilots	Tugs First Narrows
Tankers LOA < 185m < 40,000 SDWT	Yes	< 1.0k	< 0.5k	T	1	-
Tankers LOA > 185m > 40,000 SDWT	No	< 1.0k	< 0.5k	T	2	T

T = Tethered Tug Required

Table 4: Second Narrows MRA (MRA-2) deep sea vessel – tug and bollard pull requirements matrix

LOA less than 200 m and moulded breadth less than 35 m

Vessel Draft	No. of Tugs	Bollard Pull Tonnes	No. of Tugs	Bollard Pull Tonnes
	Bow		Stern	
< 8m	1	20	1	30
> 8m < 10m	1	30	1	40
> 10m	1	30	1	50

LOA 200 to 229.9 m and moulded breadth less than 35 m

Vessel Draft	No. of Tugs	Bollard Pull Tonnes	No. of Tugs	Bollard Pull Tonnes
	Bow		Stern	
< 8m	1	30	1	50
> 8m < 10m	1	60	1	65
> 10m < 12m	1 or 2	60	1	80
> 12m	1 or 2	60	1	110

LOA 230 to 250 m and moulded breadth less than 45 m

Vessel Draft	No. of Tugs	Bollard Pull Tonnes	No. of Tugs	Bollard Pull Tonnes
	Bow		Stern	
< 10m	1 or 2	60	1 or 2	65
> 10m < 12m	1 or 2	60	1 or 2	80
> 12m	1 or 2	60	2	110

As well as the rule extract cited above, the HPP standard defines requirements for visibility, rights of way, shipping priorities, and numerous other logical safeguards for shipping.

So how does this comprehensive and very current (2017) standard measure up against the "state of the art" of tanker shipping safety? If one looks first at the overall procedures, they are quite thorough and reflect a great deal of input from many stakeholders. The results are generally empirical and reflect many years of experience from all sectors of the local maritime community. It reflects effectively what is known to work well, but is it enough in light of the heightened sensitivities omnipresent today re tanker shipping? Will the precautions called for stand the test of science?

Consider the escort tug requirements within the HPP:

- The requirements only call for tethered "Tractor/ASD tugs" defined erroneously as "*... a tug with either Z-drive (or azimuth thruster) or Azimuth Stern Drive (ASD) propulsion systems capable of generating all directional propulsions forces.*" This definition confuses the type of drives used with the geometry of drive installation and leaves no room for highly effective escort tugs with Voith-Schneider propellers or with novel propulsion geometry such as a Rotortug® or a ***RAVE*** tug. More accurate definitions are required

- No formal Class approved escort rating is stipulated; only the bollard pull, a value of little use at higher transit speeds. The line force which can be generated at the stipulated maximum transit speed of 6 knots is the operative performance parameter, and must be defined
- No stability requirements for the escort tugs are defined. Transport Canada is embarrassingly silent on this matter in their outdated regulations, not addressing the potential for externally applied forces on tugs whatsoever. This is a serious regulatory omission. The recently introduced international stability standard for tugs [26] should be invoked as a minimum requirement, without reference to any Canadian regulations as alternatives; there simply are no viable Canadian regulations for this type of vessel
- The HPP is silent on the type of winch and towline to be found on the escort tugs. The line strength must be commensurate with the duty, and the winch must be designed to sustain the required line force without use of the winch brake. This criteria is an integral part of all Class requirements for escort duty and is essential to the safety of the tug should a situation go awry. It is also vitally essential to the safety of the tanker because if the tug's towline breaks due to a sudden surge load, then the critical connection to the ship is lost. The "weak link" in any escort system must be the winch brake setting and not any structural component or the towline itself. Maintaining control over the tanker throughout any escort operation is of paramount importance. Again the new BV rules provide useful guidance.

Next, consider the defined transit speed of 6 knots:

- The HPP does not mention how this speed originated, but presumably it is a speed found by ship masters and the BC Coast Pilots themselves to be one that affords an appropriate degree of control over the ship and which also ensures that the ship can be accurately and safely controlled by attending tugs
- From a tug escort perspective this speed is rather "middle ground"; it is a bit fast for effective direct thrust operation, and too slow for effective use of the indirect mode. ASD type tugs are generally most efficient in transverse arrest mode at this speed

It can be seen from the previous discussions regarding tanker resistance to damage that controlling speed of both struck and striking ship can have a significant impact on the damage results. Angle of impact is critically important; it is unlikely that any double hull structure will resist a collision at full speed with angle of incidence of 90° without significant oil outflow. However it also virtually impossible within the geography of the Port of Vancouver, given the HPP requirements, that any ship of appreciable size could be operating at right angles to the primary tanker route.

Similarly for allision, reducing the speed of passing traffic will ensure that penetration of the double hull can be avoided.

The isolation of tanker traffic within Vancouver harbour limits is an extremely effective means of mitigating risk of collision. Coupled with the speed limits cited and the use of multiple escort tugs it is more than reasonable to conclude that a spill from a tanker within the harbour limits is almost impossible, if operated under current guidelines. In the outer reaches of the tanker passage to the open sea tanker escorts have been the norm for several years and are well-practised, and utilize well-found escort tugs. The proposed expansion of tanker traffic will result in the use of faster and larger rescue tugs (untethered) in the stretch of coast through Juan de Fuca Strait to off Cape Flattery. There is ample sea room in these waters such that tethered escorts are not necessary. One can therefore conclude that although slightly imperfect, the current procedures and plans will result in very safe transit of tankers through the Port of Vancouver and adjacent coastal waters.

It is noteworthy that other efforts are underway within BC to expand the understanding of all involved with respect to tanker escort operations. The Pacific Pilotage Authority has recently published a "Discussion Document on Escort/Rescue Tug Requirements on the West Coast of Canada" [27]. This is broader in scope than just port operations. It aims to address the potential for much expanded tanker operations and gas carrier operations on the west coast of Canada, and the need for suitably powered escort tugs and rescue tugs as a major mitigating factor against a grounding and potential oil spill. This is an important document however, as it clearly addresses the need for preventative measures rather than response measures as the single most effective way to prevent any oil spill. This report correctly starts out by basing projected tug force generation capability as a function of the attended ship size, speed and rudder angle. Such data is published by the Nautical Institute (Hensen [28]), but that study unfortunately refers to required Bollard Pull and does not correlate that to line forces required at the various speeds. The numbers will vary considerably as a function of the tug hull design and its stability characteristics. Any "brick" in the water can generate a stipulated BP, but only an extremely well-designed escort tug can safely develop and sustain line pulls comparable to or ideally greater than that same BP value at speeds of 10 knots.

8. SUMMARY

Since the almost global introduction of mandated tanker escort systems in the mid-90's, there have been very significant advancements in both the design and performance of escort tugboats. In addition during that same period there has been a general improvement within the towing and shipping industries in understanding how these powerful vessels can operate most efficiently. This includes better understanding of safer escort operations and the training and familiarization required by all personnel operating within the system to ensure the safest possible operations.

Tanker escort tugs today are large, powerful, very high performance tugboats. As they are generally purpose-designed for the demanding escort role they may not always be the best "multi-purpose" tug, however it is very common for escort tugs to be fully equipped for both fire-fighting and spill response. It is important to note however that the best escort tug may NOT be the best rescue tug, as the towing gear requirements and associated deck layouts are not the same and may not be entirely compatible. On larger tugs it is easier to fit both systems, but such large tugs may not be necessary. The requirements for both tasks must be carefully evaluated and then judgments made as to whether or not a single or different tug types are necessary for effective coastal protection.

It is likely that improvements in the resistance of double hull structures to collision, allision and grounding can be made by, for example, increasing the thickness of double bottom plating. However these structural additions would come at a price, heavier ships require more power or can carry a reduced cargo thereby increasing emissions per tonne carried and also costing more.

The studies reported on in this paper show that the current rules governing construction provide effective resistance to oil outflow but only in tandem with mitigating efforts such as speed reduction, avoidance of close quarters situations and the use of suitable well-designed escort tugs.

It should be noted that the referenced studies all broadly assume that the ship remains in its "as built" condition for its service life. There have been many studies on corrosion degradation of ships structures and while it would be relatively straightforward to model the impact on damage resistance of corrosion, by simply making the structure thinner, the authors of this paper are not aware of any explicit studies.

This leads to the inescapable conclusion that for the purposes of ensuring adequate resistance to oil outflow following impact of any type the hull structure must be maintained in a near "as new" condition. Fortunately the existence of the CAP (Condition Assessment Program) scheme, a voluntary scheme, although effectively almost mandatory, ensures that tankers over fifteen years old are maintained at, or very close to, the new build standard. In this respect it is concluded that oil tankers greater than fifteen years old trading into Vancouver harbour should be maintained to a standard of CAP2 or higher. While this is a voluntary standard and so cannot, under current arrangements, be enforced by Transport Canada, it can be enforced by terminals and charterers. The fifteen year cut off is important as experience tells us that corrosion is rarely significant until after fifteen and can be controlled effectively by the application of efficient coating repairs at Year 15 when the ship is in dock for routine dry docking.

An effective in-service inspection program operated by the ship owners themselves with regularly scheduled inspections by class and oversight from oil majors will ensure that structure remains effective throughout the life of a ship. The inspection routine can again be monitored by oil majors and flag state and it is known that some owners of conventional tankers have implemented a six monthly ballast tank inspection regime. Oilcos responsible for terminals can, and do, insist on seeing inspection records, both from Classification Societies and crew before granting terminal clearance and loading cargo. These audits are carried out each and every time a ship is to be cleared for cargo.

Based on the evidence available today, it is very clear that oil tanker movements can be made as safe as any type of shipping, reducing risk of spills to almost zero. This can be accomplished however ONLY by following a carefully developed Tanker Shipping Control Plan, which incorporates at least the following measures:

- Strict controls over all shipping in the vicinity of tanker movements to give priority and right of way to the tanker and their escorts
- Limits on the environmental conditions under which tanker movements can take place
- Limits on tanker movements to daylight hours and clear visibility conditions
- Limiting tanker speed to that at which any potential collision or allision (however unlikely) would not result in a breach of a cargo tank
- Analysis of the tug forces required to effect safe control over the ship (in terms of steering or braking in whatever is the most advantageous direction) at the limiting escort speed(s), OR as an absolute minimum, stipulating tug escort forces corresponding to those which could be generated by the size of tanker itself
- Use of properly certified escort tugboats which have:
    - the defined steering and braking force capability at the required speed
    - Class-approved stability certificates which correspond to the defined loads

  - towlines in good condition which have breaking strength ratings well above (minimum 3x) the rated line forces
  - hawser winches which can carry the rated line load on a "dynamic brake" rather than on a static or manual brake which cannot release in a controlled manner under load
- Use of properly trained personnel in every aspect of the escort operation
- A system which routinely practises all possible escort response strategies to ensure crew understanding and familiarization

9. CONCLUSION

There is undoubtedly sufficient good information in the technical arena to demonstrate clearly that double hull tanker structures themselves have adequate collision and grounding resistance but only in conjunction with other mitigating measures such as navigation, speed restrictions and escort towing. Some of the work would benefit from updating given the approximately twenty five years of double hull tanker operations, since the delivery of ***Eleo Maersk*** in 1993 and the increasingly high structural standards imposed by Classification Societies. Not only have structural standards increased but significant advances in escort tug design together with traffic management also contribute to high standards of tanker safety.

10. REFERENCES

1. The Oil Pollution Act (OPA) of 1990.

2. MARPOL 73/78 is the International Convention for the Prevention of Pollution from Ships, 1973 as modified by the Protocol of 1978 and more specifically, MARPOL > Annex I - Regulations for the Prevention of Pollution by Oil > Regulation 19 – Double Hull and Double Bottom Requirements for Oil Tankers Delivered on or after 6 July 1996.

3. DECOLA, ELISE, 'OIL SPILL PREVENTION and RESPONSE IN THE U.S. ARCTIC: Unexamined Risks, Unacceptable Consequences', *Nuka Research and Planning Group, LLC Pearson Consulting, LLC*, 2010.

4. ROSS, CAPT. KEN H., 'Harbour Towage in the 90's', *Proceedings of the 12th International Tug & Salvage Convention, TRP Magazines Ltd., London, UK.*, 1992.

5. BANKS, G., 'Escort Tugs Performance Comparisons', *International Tug & Salvage Convention*, 1996.

6. LAIBLE, D.H. and GRAY, D.L., 'Escort Tug Computer Simulations – Standards and Advances', *International Tug & Salvage Convention*, 1996.

7. ROKSTAD, J. and AMUNDSEN, C.J., 'Optimised Escort Tug for Norsk Hydro's Sture Terminal', *International Tug & Salvage Convention*, 2000.

8. ALLAN, R.G. and MOLYNEUX, D., 'The Development of a New Generation of High-Performance Escort Tug', *International Tug & Salvage Convention*, 2000.

9. MINORSKY, V.U., 'An analysis of ship collision with reference to protection of nuclear power ships', *Journal of Ship Research, Vol.3, pp.1-4.*, 1959.

10. PAIK, J.K. et al., 'On Rational Design of Double Hull Tanker Structures Against Collision', *SNAME Transactions, Vol. 107, pp. 323-363.*, 1999.

11. ARITA K.A., 'Study on the collision strength of ships', *Ph.D. Thesis, Osaka, Japan, Osaka University*, 1986.

12. KIERKEGAARD, H., 'Ship Collisions with Icebergs', *Ph.D. Thesis, Denmark Technical University*, 1993.

13. WANG, G., 'Structural Analysis of Ships' Collision and Grounding', *Ph.D. Thesis, University of Tokyo, Tokyo, Japan*, 1995.

14. LEE, T.K., cited in *Journal of Ship Research, Volume 45, Number 2, pp. 111-132(22)*, June 2001.

15. SIMONSEN, BO CERUP, 'Marine Structures Ship Grounding on Rock', *I. Theory, Volume 10, Issue 7, Pages 519-556*, August 1997.

16. PAIK, JEOM KEE, 'On Collision Strength of Ship's Bow', *International Society of Offshore and Polar Engineers: The Seventh International Offshore and Polar Engineering Conference, Honolulu, Hawaii, USA*, 25–30 May 1997.

17. ZHANG, SHENGMING On Collision Strength of Ship's Bow', *DTU*, January 1999

18. BROWN, ALAN et al, 'Structural Design and Response in Collision and Grounding', *SNAME Transactions, Vol. 108, pp 447-473*, 2000.

19. 'Deliverable 3.4 – Integrated Methodology for Survivability and Oil Outflow Assessment Document ID Code: 3-40-RD-2006-14-01-0 Pollution Prevention & Control', *Herbert Software Solutions Europe (HSSE).*

20. HEINVEE, MARTIN and TABRI, KRISTJAN, 'Double-hull Breaching Energy in Ship-Tanker Collision', *Proceedings of the 6th International Conference on Marine Structures, Marstruct,* 2017.

21. PAIK, J.K. and SEO, J.K., 'A Method for Progressive Structural Crashworthiness Analysis under Collisions and Grounding', *Thin-Walled Structures, Vol.45, Issue 1, pp.15-23,* 2007.

22. AMMERMAN, D, 'Marine Safety Systems, Control Ballast Tanker Interactive CD', *SAND2002-3188P, Sandia National Laboratories, Albuquerque, NM,* 2002.

23 AARTS, SVEN, 'The development of a Class of Terminal Tugs', *Proceedings of the Tenth International Tug Convention; Sydney Australia; Thomas Reed Publications Limited,* 1989.

24. VAN DER LAAN, M, 'Carrousel Tug Design.' *Proceedings of ITS 2002; 17th International Tug and Salvage Convention; Bilbao, Spain, The ABR Company Limited,* May 2002.

25. DE JONG, GIJSBERT, 'New Regulations for Towing and Anchor-Handling Vessels: The Future is Now', *Bureau Veritas, Proceedings of the 23rd International Tug, Salvage and OSV Convention and Exhibition, The ABR Company Ltd.,* 2014.

26. 'Bureau Veritas Rules for the Classification of Steel Ships', *PART E – Service Notations for Offshore Service Vessels and Tugs; Section 2-Stability,* July 2017.

27. 'Discussion Document on Escort/Rescue Tug Requirements on the West Coast of Canada', *Pacific Pilotage Authority,* July 2017.

28. HENSEN, CAPT. HANK, FNI, 'Tug Use in Port; A Practical Guide', *The Nautical Institute,* 2003.

11. AUTHORS BIOGRAPHIES

Dr. Iain Braidwood, is a Naval Architect at his own Consulting Company. Extensive experience of specification, repair and operation of both single and double hull tankers working for major ship owners and a Classification Society lends itself directly to this paper.

Robert G. Allan, P. Eng. is the Executive Chairman of Robert Allan Ltd., consulting Naval Architects based in Vancouver, BC. Recently (mostly!) retired, his role is essentially to act as mentor to a new generation of talented naval architects within the firm. He has written extensively on the subject of escort tug design and escort operations and has received several major peer awards for this work, including the RINA Small Craft Group Medal in 2005.

Damage Control Survivability & Containment of Fire, Smoke and Flooding. Only as strong as the weakest link: Multi Cable Transits (Glands) & Pipe Penetration's critical role in achieving fire and watertight integrity.

Peter J Litchfield, CSD Sealing Systems, UK

SUMMARY

The effective compartmentalisation of a ship is critical to the containment of fire, smoke and flooding to ensure survivability. The containment is only as strong as the weakest link; the sealing of the cable and pipe penetrations through the bulkhead and decks play a critical safety role but they are often not looked at closely enough to ensure they are installed correctly in strict accordance with manufacturer's guidelines and certification limitations and insulation requirements.

Recent incidents have highlighted fundamental systemic failings particularly within cable penetration transits.

To maintain the vessels original design safety intent, cable and pipe penetrations must be viewed as a safety critical system with all the associated system controls in place for initial installation and any modifications through life.
The paper explores the risks, the potential failure mechanisms and the quality control measures that can be introduced to improve the safety.

NOMENCLATURE

ANEP	Allied Naval Engineering Publication
CEFOR	The Nordic association of Marine Insurers
DMAIB	Danish Maritime Accident Investigation Board
FTP	Fire Test Procedures
IMO	International Maritime Organisation
SOLAS	Safety of Life at Sea
UT	Ultrasonic Technology

1. INTRODUCTION

The effective compartmentalisation of a ship is critical to the containment of fire, smoke and flooding to ensure survivability. The general requirement of the IMO concerning design of ships and offshore installations is to 'contain a fire in the space of origin.'

The penetrations must maintain the strength, water-tightness and fire resistance of the vessel's structure and so act as a critical safety barrier contributing to the overall safety of the vessel. It is therefore essential that each individual penetration sealing system installed is fit for its intended purpose.

Type approved penetration sealing systems are generally used, as they carry evidence of being designed and tested to ensure fireproof, watertight and/or gastight penetrations by cables and pipes in bulkheads and decks.

Functional requirements for penetrations are detailed in Naval Ship Code and SOLAS;

- SOLAS Chapter II-2 Regulation 9.3 - Penetrations in fire-resisting divisions and prevention of heat transmission [1]
- SOLAS Chapter II-1 Regulation 13 and 13-1 for passenger- and cargo vessels respectively. [2]

Detailed design and test requirements for penetrations are specified in the IMO Fire Test Procedures (FTP) Code 2010 [3] as well as in rules set out by Classification Societies.
The suitability of each penetration and sealing system is subject to approval and inspection by Classification Societies and/or Flag States as part of a vessel's safety certificates.

For ship stability calculations there is a general assumption that the watertight bulkheads and hence the penetrations will remain tight and contain the flood and therefore most calculations assume no progressive flooding occurs through penetrations.

Following Investigation of several engine room flooding incidents CEFOR- (The Nordic Association of Marine Insurers) published a report in May 2016 highlighting the failings of incorrectly installed and poorly maintained boundary cable penetration sealing arrangements contributing to the development and seriousness of accidents.

During the ship design & build process, focus is on the steel construction of the watertight bulkheads entailing quality control of the design and qualifications of the welders and testing of the welding and thickness measurements of the steel. The cable penetration are viewed from a component viewpoint without anywhere near as much independent quality control or testing and with separate approval procedures that often do not involve testing in situ and therefore performance depends on the individual craftsman assembling the system.

There is often no readily available, in-situ, testing method for assuring the quality of the fitted cable penetration system.

The classification survey process is not always performed by a person with detailed knowledge of the cable penetration system and therefore, process rests on the type approval certificate.

The FTP Code recognises that penetrations are a local weakness and details rigorous test requirements however there is often a lack of control to ensure penetrations are installed as tested and in strict accordance with manufacturers specific instructions.

Recent reports on fire & flooding incidents have highlighted some of the failure mechanisms however additional factors need to be considered including the through-life performance of cable and pipe penetrations which is key to survivability during fire or flooding.

One recently identified survey method has been the use of ultrasound for watertight integrity surveys. Whilst ultrasound provides an additional level of assurance it will not pick up all the potential failure mechanisms and therefore should be used in combination with an experienced installer or surveyor.

2. CONTAINMENT OF FIRE

The longer a fire can be contained in the space of origin, the greater the chance of survival. A ship is therefore compartmentalised and the fire performance of the divisions are determined in accordance with SOLAS Chapter II-2 Regulation 9 - Containment of Fire (or ANEP 77 Naval Ship Code) which prescribes the minimum fire integrity and insulation of all bulkheads and decks depending on the fire hazard and risk of the adjacent compartments.

Many of the divisions are rated as "A" class divisions; either "A-60" or "A-0", both designed to contain a fire for 1 hour.

In addition to the vessel's Insulation Key Plan there also should be a Fire Classification Plan (Figure 1) that details all the fire boundaries including the non-insulated "A-0" and "B-0" fire boundaries.

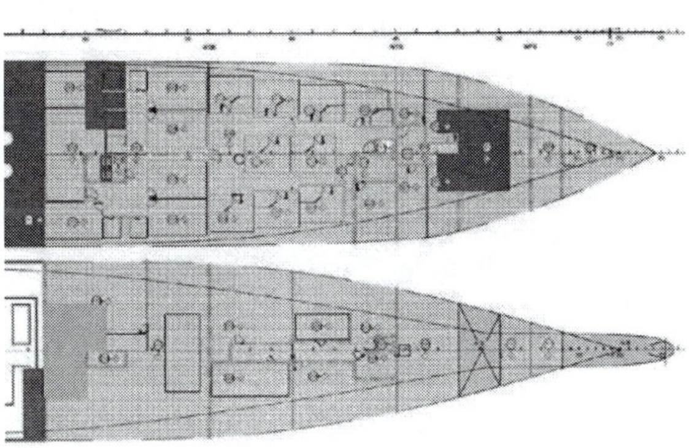

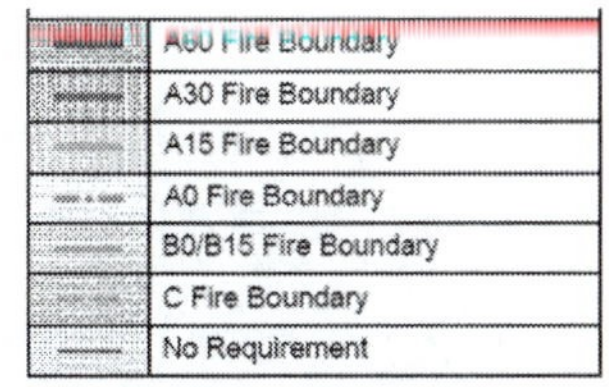

Figure 1: Typical Fire Classification Plan of Bulkheads and Deckheads

2.1 SOLAS DEFINITION "A" CLASS FIRE DIVISIONS [4]

It is important that the SOLAS definition of "A-0" boundaries are correctly understood.

"A" class divisions are those divisions formed by bulkheads and decks which comply with the following criteria:

- they are constructed of steel or other equivalent material;
- they are suitably stiffened;
- they are insulated with approved non-combustible materials such that the average temperature of the unexposed side will not rise more than 140°C above the original temperature, nor will the temperature, at any one point, including any joint, rise more than 180°C above the original temperature, within the time listed below:

class "A-60"	60 min
class "A-30"	30 min
class "A-15"	15 min
class "A-0"	**0 min**

- **they are constructed as to be capable of preventing the passage of smoke and flame to the end of the one-hour standard fire test;**

2.2 PENETRATIONS THROUGH "A" CLASS FIRE DIVISIONS

"A" class fire divisions may have to be provided with apertures to allow them to be penetrated by cables and pipes and it is necessary to reinstate the insulation and integrity performance of the division at the position where it has been penetrated. [2]

The FTP code has always recognised that pipe and cable penetrations are a local weakness in any division and requires that penetration sealing systems are independently fire tested and type approved [3]. However, type approval alone does not guarantee performance.

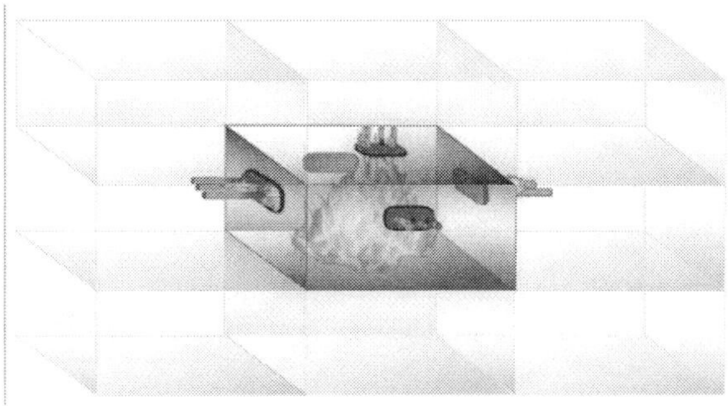

Figure 2: Rapid spread of fire to multiple compartments through penetrations.

If the cable transits or pipe penetrations fail (Figure 2), fire and smoke can spread rapidly to multiple compartments, and very quickly a situation can get out of hand, especially if there are multiple failures in various locations.

The correct installation, as fire tested, including any additional insulation applied is critical to the effective fire performance of pipe penetrations and multi-cable transits.

Many penetration systems require additional insulation added to achieve the required "A" class fire performance tests both for "A-0" and "A-60" particularly if the fire hazard is from the non-insulated side of the bulkhead.

This insulation therefore becomes an integral part of the system and is normally referenced in the type approval certificate drawings but unfortunately this insulation is often over looked and not applied in practice which can result in the failure of the cable transit and the spread of fire.

2.3 PENETRATIONS THROUGH "A-60" DIVISIONS

There are problems with the way cable transits and pipe penetrations are insulated in "A-60" fire divisions:

For penetrations in "A-60" Divisions in addition to preventing the passage of smoke and flame they must prevent temperature at any point rising more than 180 ^{0}C above the original temperature for 60 minutes [4].
For "A-60" class bulkhead fire tests the FTP code requires the penetrations to be fire tested with the fire from the normally uninsulated side of the bulkhead, representing the worst-case scenario with the fire directly against the steel, as there is often a fire hazard from either side of a fire boundary, but in order to pass the fire test and protect the penetration many cable transits and pipe penetration systems are fitted with additional insulation on this normally uninsulated fire hazard side. This insulation therefore becomes an integral part of the penetration system but is often not applied in practice.

This additional insulation detail is normally contained in drawings referenced in the type approval certification, but the drawings are often not obvious from the certification and again this insulation is often missed in practice, resulting in premature failure of the penetration.

This is a typical detail showing the additional insulation required, note:

- Insulation is extended to cover the face of the transit by at least 20mm.
- Additional insulation added on normally uninsulated side

Figure 3: Additional A60 insulation often required.

2.4 PENETRATIONS THROUGH "A-0" FIRE DIVISIONS

The term "A-0" is misleading and causes confusion, with many in the industry wrongly assuming "A-0" means no fire. Contrary to general opinion **all** "A" class divisions including "A-0" must prevent the passage of smoke and flame for one hour [4] thus preventing the escalation of the fire. The "0" references there is no requirement to maintain the temperature below 180^{0}C.

Most uninsulated steel divisions are classified as "A-0" and whilst most uninsulated steel divisions will provide integrity for one hour ("A-0") the penetrations themselves often require additional insulation to achieve the "A-0" requirement; to prevent the passage of smoke and flame to the end of one-hour standard fire test.

"A-0" conditions can be more severe than "A-60" for cable transits and pipe penetrations because there is normally no insulation protecting the transit and therefore the transit is subjected to a higher heat load because of the extra conducted and radiated heat particularly when transits are large and heavily filled with large copper cables.

The new 2010 FTP Code recognises the significant risk of integrity failure (spread of fire through the passage of smoke and flame) resulting from some manufacturers uninsulated cable transits and pipe penetrations within "A-0" uninsulated divisions [3];

- "A-0" class cable transits and pipe penetrations are recommended to be performed in an uninsulated ("A-0") bulkhead/deck. If the cable transits are tested as "A-60" penetration, any insulation fitted on an exposed side (on the cable transits itself and 200 mm around) will be required to be fitted also for "A-0".

- "A-0" cable transits shall not be approved without an "A-0" test although tested and approved as "A-60".

Many cable transits therefore require additional insulation on both fire hazard sides to protect the transit to pass an "A-0" test. The Type approval certification references drawings which detail the tested insulation arrangement for "A-0." Unfortunately, in practice this is not always applied in the field and can result in premature failure.

It is important to also consider what was actually fire tested:

- Any additional insulation added
- The maximum size of penetration tested
- Size and types of cable tested
- The cable fill cross sectional area

3. CONTAINMENT OF FLOODING

The longer a flood can be contained in the space of origin, the greater the chance of survival.
Failure of penetrations can lead to progressive flooding (Figure 3) therefore the performance of the cable transits & pipe penetrations is critical.

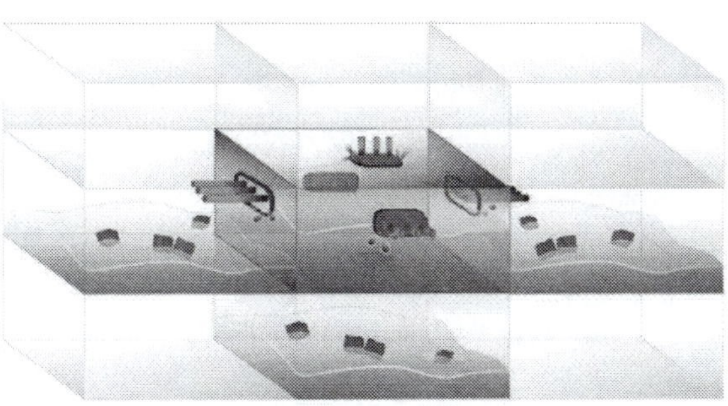

Figure 4: Progressive flooding through penetrations

- Correct installation of Cable & Pipe Penetrations, as tested and approved, is also critical to the effective flood and gas tight performance.

Type approval alone does not assure performance through life as highlighted in recent flooding cases:

4. INVESTIGATION OF RECENT ENGINE ROOM FLOODING CASES

4.1 INVESTIGATIONS BY CEFOR [5]

Investigations by CEFOR, the Nordic association of marine insurers, of a number of Engine room flooding cases revealed that incorrectly installed boundary cable penetration sealing arrangements contribute to the development and seriousness of accidents.

The technical memo issued emphasises the risks associated with type approved cable penetrations when seal packing systems are not installed as per makers instructions nor subsequently sufficiently verified by the many parties involved; i.e. Flag States, Classification Societies, Owners and Shipyards.

4.1(a) Insurance Case 1: Emma Maersk Container Vessel - Built: Denmark, 2005

In February 2013, severe flooding occurred in the large container ship EMMA MÆRSK while the ship, loaded with about 14,000 containers, was about to pass through the Suez Canal.
The situation became complicated because the ship had just initiated a passage in a convoy through the Suez Canal. Loss of the ship's own propulsion, electric power, steerage and manoeuvrability could be foreseen and eventually occurred.

The main technical sequence of events were a break-down of the forward stern thruster causing a major leakage into the shaft tunnel, failure of the watertight bulkhead at frame 166 between the shaft tunnel and the engine room, primarily caused by failure of the cable penetration seals.
Four large cable penetration seals mounted in the watertight bulkhead between the shaft tunnel and the main engine room could not withstand the hydrostatic water pressure and burst open at just 0.9 bar and eventually the main engine room was also flooded.
A cable penetration to the cargo hold (Figure 5) also burst open at just 0.5 bar.

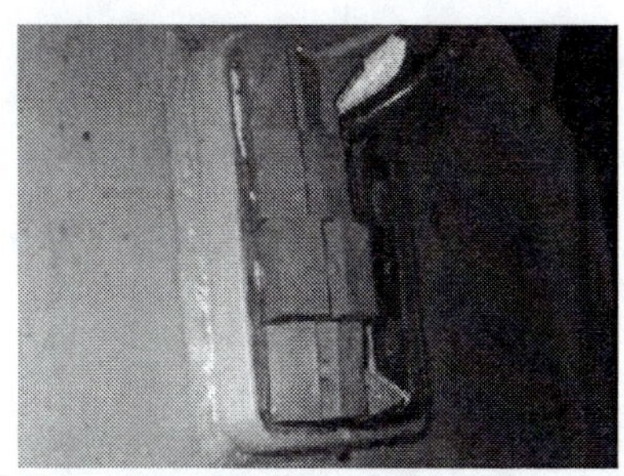

Figure 5: Emma Maersk Failure of cable penetration to cargo hold

Before the cable penetrations burst, water was seeping from the cable penetrations indicating that, besides the use of plastic stay plates instead of steel stay plates, there were other problems with the assembly of the cable penetration components.

4.1(b) Insurance Case 2: Drilling Rig - Built: China, 2012

During a yard stay in Norway, a rig built in China 2012 experienced an unfortunate damage and flooded a part of one pontoon to approx. 8 meters height.

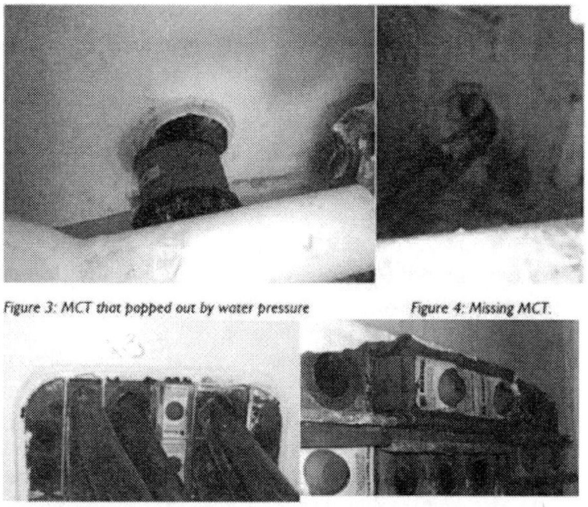
Figure 3: MCT that popped out by water pressure Figure 4: Missing MCT.
Figure 5: Visible gap between cables and module halves. Figure 6: Wedges are not tightened

Figure 6: Failure of cable penetrations on Drilling Rig.

Some of the cable penetrations did not withstand the water pressure and "popped out". Others were leaking, extending the flooded area of one pontoon and the damage to expensive equipment.
Many and different faults (with reference to makers set of criteria for correct installation) were observed.

4.1(c) Insurance Case 3: General Cargo Vessel - GT 2.973 - Re-Built: Romania, 2002

In January 2010, in a freezing cold Baltic sea, it was discovered that water ingress was filling up the vessel's engine room aft, and the vessel was in danger of sinking.
Investigations established the water ingress originated from the emergency fire pump under the fo'c'sle, where it had penetrated the first water tight bulkhead and filled up the tunnel, before it penetrated the second watertight bulkhead entering into the engine room. It was also revealed that the cable openings in the mentioned bulkheads probably had not been equipped with any water tight sealing at all when the vessel was rebuilt in 2002.

Figure 7: Bilge alarm and cable duct through the bulkhead on frame 112 – not sealed

4.1(d) Insurance Case 4: Bulk Carrier - DWT: 177 800 - Built: 2009

In 2012, the vessel grounded in loaded condition on hard bottom surface resulting in severe damages and flooding to the forward part of the double bottom including pipe tunnel.

A couple of days after the grounding, a casted cable penetration between the pipe tunnel and engine room burst and caused complete flooding of the engine room.

4.1(e) Findings from the CEFOR report were:

- Integrity of seal packing systems is a challenge across vessel types, Yards, Classification Societies and Owners.
- Not possible to conclude only on a single cause to these costly incidents, as many parties are involved over a vessel's lifetime.
- Type approval certification alone is not enough to ensure safety as one of the contributing factors seems to be lack of 'knowledge' and 'attention' to detail amongst the many parties involved in installation and verification of these seal packing systems.

4.1(f) Recommendations from CEFOR:

- Inform owners site team about the risks associated with type approved cable penetrations and emphasise the need for team members to possess the necessary competence to verify if seals are installed according to maker's requirements.
- Include check of seal packing systems in vessel's Planned Maintenance System and dry docking program.
- Enhance tests and verification processes performed by Classification Societies for both newbuilds and annual surveys.

Looking at the Emma Maersk in more detail:

4.2 EMMA MAERSK- FLOODING OF ENGINE ROOM

Findings and recommendations following the failure of the cable transits according to the report by the Danish Maritime Accident Investigation Board [6]

4.2(a) Analysis

Making the effectiveness of such a safety critical installation dependent on one craftsman reveals a weakness in the construction and approval process of the watertight bulkhead. Bearing in mind that the construction and approval of other parts of the watertight bulkhead is subject to several quality control measures, it

becomes apparent that there is less focus on the cable penetration system.
Lack of systematic training, the necessary general and skill-based knowledge to carry out an effective installation of a cable penetration was obtained by "on the job training" with the risk that important elements with regard to correct mounting were not taken into account. It was therefore up to the individual craftsman to determine penetration.

The type approval process for cable penetrations is system-based and not component-based. This means the approval is not based on a specific configuration of the cable penetration system, but dependent on correct installation, in accordance with the installation guidance, type approval drawings and any limitations

Without any test procedure after the installation, the quality of the cable penetration system is entirely dependent on the knowledge and skill of the individual craftsman.
When installed, it was difficult to identify any discrepancies from the installation guidance whereupon the approval rested. As per procedure, the on-site surveyor to the classification society based his approval of the cable penetrations of the watertight bulkhead on the type approval certificate and on a visual inspection. He had no detailed knowledge of how the cable penetration system was supposed to be fitted and he could not react to the discrepancy as the installations did not appear to be any different from other cable penetrations. Nor did the surveyors of the yard or the owners reveal that the cable penetrations did not meet the type approval criteria.
To understand why the cable penetrations in the watertight bulkhead had been mounted incorrectly and why it went unnoticed, need to look beyond the failure of the individual craftsman and address several systemic issues:

- In the design process, focus was on the steel construction of the watertight bulkhead entailing quality control of the design and testing of the welding and thickness measurements of the steel. The cable penetrations were viewed from a component viewpoint
- The cable penetration system was approved as a system and not as a single component on conditions that finally rested on the individual craftsman assembling the system.
- The craftsman did not have detailed knowledge about the status of the bulkhead as a whole.
- The classification survey process was not performed by a person with detailed knowledge of the cable penetration system. Therefore, the survey process rested on the type approval certificate.
- The surveyors of the yard or the owners did not reveal that the cable penetrations did not meet the type approval criteria.

4.2(b) DMAIB Conclusions

The watertight bulkhead did not function according to the designed intentions as a result of the gap in the continuity between the construction and approval process. There will inevitably be some difference between a design and its construction. Any weaknesses in these discrepancies can sometimes be revealed by a survey.
The survey process has its limitations, primarily because the surveyors do not necessarily have any detailed knowledge about the mounting of the cable penetration system and consequently rely on the type approval of the system. Therefore, the structural integrity of the watertight bulkhead rested solely on the craftsmen assembling the cable penetration system, and they may not have been aware of the safety critical status of that particular penetration as part of a watertight installation. Any mistake made during the mounting of the cable penetration would, in the given circumstances, remain unnoticed.

4.2(c) Actions Taken & Preventative measures

4.2(c)(i) Classification Society

American Bureau of Shipping has informed:
"Steel Vessel Rules [see 4-8-1/ 5.3.1], maintain a longstanding requirement for a 'Booklet of Standard Wiring Practices' to be submitted for approval by builder before proceeding with the work. For cable penetrations through watertight, gastight and fire rated bulkheads and decks, evidence of penetration design approval is to be submitted. For watertight and gastight cable penetration approval, certificates issued by a competent independent testing laboratory would be acceptable.
Alternatively, hydrostatic and pneumatic pressure testing of penetrations carried out at the manufacturers' facility and witnessed by an ABS Surveyor or other competent third party would be accepted.

Additional Preventative Measures:
2013 July 1st – ABS Steel Vessel Rules [see 1-1-7/3] Machinery Plans were amended to require that 'Ship Equipment List' include "Where electrical cables, hydraulic lines, etc., penetrate watertight or fire rated bulkheads by the use of standardized penetration kits, a schedule is to be provided indicating the location, number, manufacturer, model number and type of Bulkhead Penetration Devices provided to maintain the bulkhead integrity."

Corresponding 'Initial New Construction Survey tasks' (questions) exist as Nos. 175, 363 & 491 on surveyors checklist with similar intent / text as above.

2013 July 1st – ABS Steel Vessel Rules [see 7-6-2/1.1.8] were amended to by adding "1.1.8(b) Cable Penetrations (1 July 2013). Watertight and fire-rated cable penetrations in decks and bulkheads to be generally examined for alterations and continued effectiveness."

An 'Annual Machinery Survey' checksheet task (no 15) has been created to insure that the intent of the rules is carried out.

2013 August 6th - ABS Chief Surveyor, Mr. L.A. Pendexter, issued a 'Surveyor Awareness' letter to surveyors worldwide listing changes to the Steel Vessel Rules with specifics.
ABS places significant emphasis on watertight & firetight boundaries in accordance with ABS Steel Vessel Rules and international conventions. Aforementioned changes have been effected to strengthen awareness and adopt additional measures as part of new construction and after construction survey process."

4.2(c)(i) The Owners

Based on the findings on Emma Maersk:

- Sister Vessels have been checked and material replaced in critical bulkheads accordingly.
- All Maersk Line operated vessels have inspected cable penetrations mounted in water tight bulkheads in accordance with Damage Control Booklet/Damage Control Plan.
- A test of cable penetrations in watertight bulkheads has been carried out in accordance with revised ABS requirements. Record of the result has been entered in a logbook with location (frame number/deck) for presenting to the attending surveyors.
- The ongoing new building projects work instructions or site supervision have been enhanced.
- Meetings with Class, Yard and Makers on site have been carried out in order to confirm and ensure correct material specification/certification as well as correct installation method.
- Inspections of cable penetrations have been added to planned maintenance program carried out on all Maersk Line operated vessels.

5 WATERTIGHT INTEGRITY THROUGH LIFE

Before the Emma Maersk cable penetrations burst, water was seeping from the cable penetrations indicating that, besides the use of plastic stay plates instead of steel stay plates, there were other problems with the assembly of the cable penetration components [6]. The plastic stay plates were rated to 1 bar and yet failed at less than 0.9 bar and 0.5 bar.

5.1 TYPE APPROVAL PRESSURE TEST

For Type Approval, manufacturers generally invite class surveyor or other competent 3rd party to witness a hydrostatic and pneumatic pressure testing of penetrations soon after the seals have been installed in the laboratory. There is currently no requirement to determine the performance through life by pressure testing a sample after a few months or years.

5.2 LOSS OF SEALING FORCE THROUGH LIFE

"The loss of the sealing force during the life of a seal under compression is well known (Bunting et al, 1992). Compression set and compressive stress relaxation are normally the key issues when considering the performance of static seals. [Coveney & Rizk (2005) and Morgan et al (2005) consider other causes of failure.] There will in the seal's life come a point when the retained sealing force is inadequate to overcome the leakage pressure of the fluid, and hence the life of the seal is at an end." []

"water leakage past a seal is possible when the water pressure exceeds the contact pressure exerted by the seal" [8].

5.3 COMPRESSION STRESS RELAXATION LOSS

Compression stress-relaxation is considered to be the most direct measure of the sealing capacity of an elastomer. It gives the force a rubber article would exert on a solid sealing surface as a function of time. This is only an issue for penetration sealing systems that use compression to create a seal and is not an issue for systems that use adhesive forces to create the seal.

A compression stress relaxation loss of 50% is considered significant and could result in failure of the seal.

The results of testing by TNO Laboratories on rubber used in compression designed multi-cable transit penetration seals show a significant loss of greater than 50% in the compressive force, within just 2.5 months to 2 years indicating there will be a significant reduction in the sealing performance.

Compression stress relaxation loss of typical rubber used in penetration seals		
	2.5 months	2 years
Rubber Type 1	60%	72%
Rubber Type 2	45%	56%

The effect of compression stress relaxation of the cable sheath elastomer will also be a factor that will also reduce the sealing force.

5.4 COMPRESSION SET

Compression Set is a measure of the ability of a material to recover and resist permanent deformation after being compressed.

As a rubber is compressed over time, the sealing performance decreases as rubber and cable sheath become permanently compressed and loses ability to return to original thickness.

One manufacturer recognises this problem and states within their installation procedures; "For pressurised applications, all components must be replaced with new material after removal and refitting." On a recent project they have clarified this includes all components in and above the row where a new cable is added and includes the compression wedge. This is even after just 3 months of installation.

Unfortunately, many installers, owners and class surveyors are not aware of this requirement and therefore integrity could be compromised.

This comes at considerable cost and more importantly how can you control if they have been replaced?

The Royal Navy Type 23 COM and Hunt COM BR 's recognise that in order to achieve a watertight seal any transit seals that have been disturbed within the system when adding or changing a cable must be replaced with new components.

6 HOW DO WE IMPROVE THIS POSITION?

6.1 INSTALLATION

Correct installation during initial construction is essential to ensure the penetrations will perform as a critical safety barrier.
Only suitably competent, well informed, skilled personnel and supervisors with adequate training should install penetration systems in watertight and fire-rated bulkheads.
Ideally these personnel should be 3rd party certified.

A log should be kept of all penetrations including details of the company installing, the installers name, date installed and ideally photographic evidence of correct installation. This will help ensure installers are accountable and take ownership and responsibility for their work.

6.2 TESTING

Functional tests can prove difficult to carry out on structural barriers such as watertight bulkheads.

6.2(a) ULTRASOUND TESTING

'Ultrasonic watertight integrity testing' is capable of testing all types of weather, water and air tight seals where an operator can gain access to both sides of the seal [9].

General UT equipment consists of two main units; a generator and a receiver. The generator produces a ultrasonic modulated tone, usually at a round 40kHz which is positioned on one side of the seal. The receiver is then used by the operator from the far side of the closed seal. If at any point the seal is imperfect, the ultrasonic signal will pass though the leak in the seal, which can then be detected by the receiver.

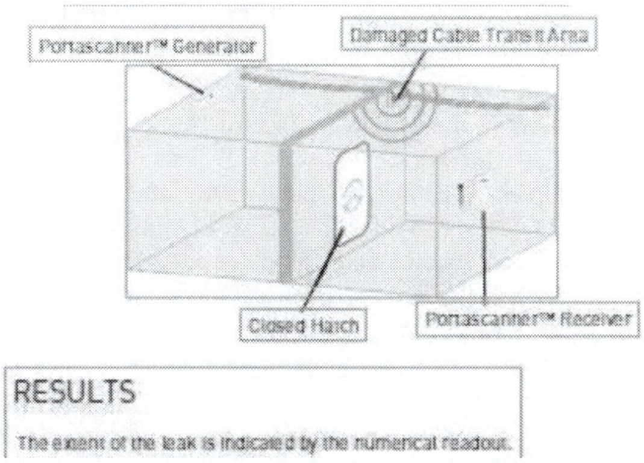

Fig. 8 How an ultrasonic signal can be used to detect the presence and location of a leaking penetration

Due to its convenience, ultrasonic tests can be conducted more frequently and can contribute to safety management and preventative maintenance procedures on board, Significant safety critical seals which might normally be ignored in a maintenance schedule can be inspected in a matter of seconds, and the exact location and source of an issue pinpointed for the operator. This offers a dramatic increase in the flexibility and ease of testing which in turn, enables seal testing to be included into a regular maintenance schedule.

The UK Royal Navy are using Ultrasonic Technology to identify any cable penetrations which are incorrectly or incompletely installed and allowing for remedial action to be taken. Across the fleet of Minehunters a very modest maintenance program has allowed for a dramatic reduction in significant defects being detected, resulting in a structural improvement across the fleet.

Recent survey experience with the RN Type 23 destroyers has also revealed that UT should be combined with an experienced detailed visual inspection to identify other defects which could still lead to failure under a hydrostatic pressure;

- Missing stay plates
- Blocks protruding beyond stay plates; blocks must be captivated by stay plates or they can be pushed out under pressure.
- Mixture of different manufacturers blocks
- Inadequate tightening of compression wedge
- Cables not entering perpendicular

6.3 SURVEY & INSPECTION

Classification Society surveyors check the watertight & fire resistance of all bulkheads and decks, including doors and penetrations but they often don't have detailed knowledge or time available to thoroughly check that penetrations are installed correctly in accordance with installation instructions and Type Approval certification;

- type approval drawings showing insulation detail
- certification limitations.

Evidence of type approval certificate and visual identification alone does not ensure the systems have been correctly installed.

Qualified, trained and experienced teams, such as at CSD Sealing Systems provide specialist training & survey inspections of cable and pipe penetration systems including the use of Ultrasound techniques. Their services are inspected and certified by Lloyds Register QA through ISO 9001:2015 accreditation.

CSD Sealing Systems are the UK representative of Beele Engineering, a leading company with over 40 years' experience in the development, production and worldwide supply of penetration seals.

To help resolve concerns over installation standards Beele have formed a partnership with KIWA (a Dutch Notified Body) to develop a 3rd party approval for all parties involved with the installation and inspection of cable transits and pipe penetrations.

7 SAFE RETURN TO PORT

Since July 2010 SOLAS requires that passenger ships of minimum 120m in length or having three or more main vertical zones must be able to return safely to port under their own power and to provide passengers with safe areas on board, after a pre-defined flooding or fire incident [10]. Placing even greater reliance on the performance of the cable and pipe penetrations within fire & watertight bulkheads and decks to contain the fire or flood.

There should be a requirement for all cable and pipe penetrations to be assessed as a **critical system** in accordance with 06/2010 MSC.1 Circ 1369 Interim explanatory notes for the assessment of passenger ship systems' capabilities after a fire or flooding casualty.
Requires:

- Overall assessment of all essential systems.
- Detailed assessment of critical system; if the system is identified as having a possibility to fail as consequence of fire or flooding casualty not exceeding the casualty threshold, then the system is considered as "critical".
- Description of tests, inspection and maintenance plan related to the concerned essential systems.

8 CONCLUSIONS

Watertight and fire-rated cable and pipe penetrations are an essential Safety Critical System which must have controls put in place to ensure the penetration will perform in the event of a damaged ship. They play a critical safety role in the containment of fire, smoke and flooding to ensure survivability. To maintain the vessels original design safety intent, cable and pipe penetrations must have all the associated system controls in place for initial installation and any modifications through life.

Relying on Type Approval certification alone is not enough to ensure safety. One of the contributing factors in system failures seems to be lack of 'knowledge' and 'attention' to detail amongst the many parties involved during installation and verification of these penetration sealing systems. The Type Approval process for cable penetrations is system-based and not component-based. This means the approval is not based on a specific configuration of the cable penetration system, but dependent on correct installation, in accordance with the installation guidance, type approval drawings and any limitations. Type Approval of penetrations should also include a measure of likely through life performance by pressure testing a sample a few months after installation followed by periodic testing through an expected lifetime or performing compression stress relaxation tests to determine any decrease in the system pressure rating.

Many penetration systems require additional insulation to be added to achieve the required "A" class fire performance tests both for "A-0" and "A-60" particularly if the fire hazard is from the non-insulated side of the bulkhead. This insulation therefore becomes an integral part of the system and is normally referenced in the type approval certificate drawings; unfortunately the insulation requirement is often over looked, especially in "A-0" boundaries, and not applied in practice which can result in the failure of the cable transit and the spread of fire.

Correct installation during initial construction and through life is essential to ensure the cable transits & pipe penetrations will perform as a critical safety barrier. Installation of the system must be as fire tested, including any additional insulation required by the Type Approval Certification. Only suitably competent, well informed, skilled personnel and supervisors, with adequate training, should install penetration systems in watertight and fire-rated bulkheads. These personnel should ideally be 3rd party certified. A log should be kept of all penetrations including details of the company installing, the installers name, date installed, where possible photographic evidence of correct installation and the results of any ultrasound testing. This will help ensure installers are accountable and take ownership and responsibility for their work. An Owners site team must be aware of the risks associated with type approved cable penetrations and ensure the team members possess the necessary competence to verify the seals are installed according to manufacturer's and Type Approval requirements.

Watertight and fire-rated cable penetrations in decks and bulkheads must to be periodically examined to ensure continued effectiveness. Primary focus should be on any new cable routes to identify any alterations. Ultrasound Testing offers a dramatic increase in the flexibility and ease of seal testing which in turn enables testing to be included into a vessel's Planned Maintenance System and Dry Docking Program. UT should be combined with a detailed visual inspection to identify other installation defects which may not be picked up by UT but could result in failure in the event of a flood.

9 REFERENCES

1. SOLAS 1974, Chapter II-2: Construction-Fire protection, fire detection and fire extinction.

2. SOLAS 1974, Chapter II-1: Construction-subdivision and stability, machinery and electrical installations.

3. 2010 FTP Code as per Res MSC 307(88) 01.07.2012 became mandatory, includes up-dated and new test procedures.

4. SOLAS Reg.3., 'Definitions of A class bulkheads'.

5. CEFOR Technical Forum, Sealing systems used in watertight cable penetrations, May 2016

6. Danish Marine Accident Investigation Board, Emma Maersk Flooding of engine room on 1 February 2013, December 2013.

7. M. Otsuka, T. Okamura, N. Suetsugu, T. Ohta and S.Ono, "A new concept of static rubber gasket for sealing rough surface," SAE technical paper 2003-01-0485, 2003.

8. JR Daley, Assessment of Life Prediction Methods for Elastomeric Seals, Elastomers and Components, 2006.

9. Coltraco Ultrasonics UK, "The might of the lesser-thought-of watertight, weathertight and gas-tight NBCW seals in warfare: why their improved maintenance today, and introducing continuous monitoring in the future, can improve safety on warships", RINA Warship 2016

10. Bureau Veritas, Safety of Ro-Ro Passenger & Cruise Ships', NI 388 - Revision 12, January 2018.

10. AUTHORS BIOGRAPHY

Peter Litchfield holds the current position of Technical Sales Director at CSD Sealing Systems Ltd. CSD have specialised in the supply, inspection and installation of technically advanced multi-cable transits systems and flexible pipe penetrations in the UK since 1993.

During his 20 years with CSD Peter has carried out and also been involved with many fire surveys and watertight integrity inspections whilst supporting Lloyds Register Surveyors, Ship Builders and Owners. His knowledge and experience has helped provide product solutions that improve safety and help ensure fire and watertight integrity across the Naval, Marine and Offshore sectors.

HAVE THE BENEFITS OF EMERGENCY RESPONSE SERVICES BEEN OVERLOOKED FOR NON-OIL TANKERS?

D Prentice, D Widger, Lloyd's Register EMEA, UK

SUMMARY

Enrolment with recognised emergency response service (ERS) providers was initially driven by the commercial needs of operators of various ship types. Whilst the later introduction of legislation, mainly for oil tankers, has meant that almost all oil tankers are enrolled, it has nevertheless created a mind-set that is focused on compliance and oil pollution, with ERS only being of benefit for this sector.

In practice, the benefits of ERS can be enjoyed by all ship types however; the majority of non-oil tankers are not enrolled. We look at key sectors that can also realise these benefits, including bulk-carrier, container, Ro-Ro, general cargo, LPG, passenger ships and yachts.

Whilst the typical casualty scenarios are generally known, it is considered that the complexity and level of the challenges faced are not. We ask, "Are the benefits of ERS properly appreciated?"

Almost every case is different in its own way and drawing upon Lloyd's Register's 35 years' experience of over 400 emergencies, a variety of scenarios are described, demonstrating the challenges, risks and benefits of rapid assessment in stabilising an emergency. And beyond that, the potentially very significant value of post-emergency assistance in supporting efforts to reach a best place of repair and to have the vessel back in service quickly and at minimum cost.

Demands on operator's emergency response resources may be very high and the ability of ERS providers to support this is also highlighted.

1. INTRODUCTION

Many thousands of vessels are enrolled in emergency response services (ERS) worldwide in case of casualty. Accidents happen, regardless of how well ships are designed and built; how well they are operated; and safeguards in place. Since the early 1980's ship operators from around the world have been approaching classification societies in need of emergency technical support in a crisis, driving the development of these ERS services.

Early clients were operators of all ship types, seeking support to manage the risks in a casualty.

Figure 1 illustrates the estimated number of ships enrolled by ship type with all ERS providers.

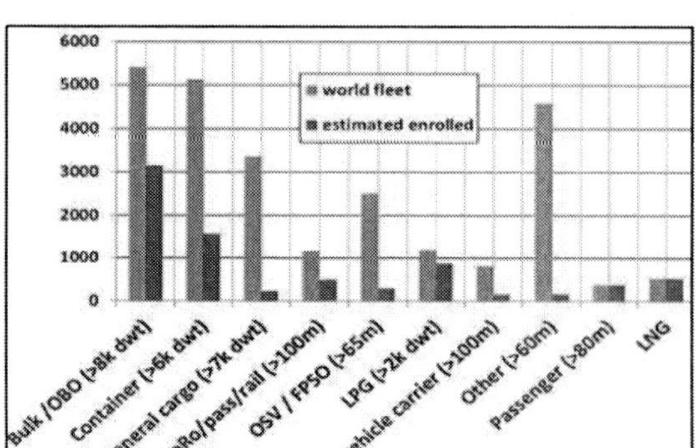

Figure 1 Main ship types – world fleet and estimated numbers enrolled with all ERS providers

With the exception of LNG and passenger ships, for most non-oil tanker ship types the majority are not enrolled, estimated at about 70%, amounting to 18,000 ships.

The casualty rate of each ship type is broadly comparable, as shown by Figure 2.

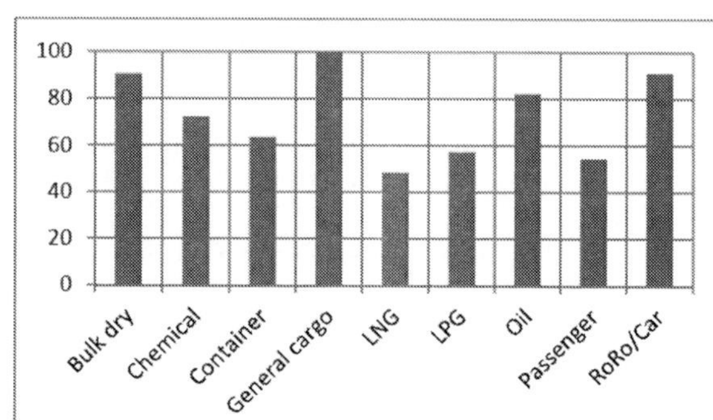

Figure 2 Comparison of ship type casualty rates (% relative to max. general cargo) (SERS)

The various benefits of ERS to the operators of these vessel types are discussed in the sections that follow.

2. CHALLENGES FACED BY OPERATORS

In case of casualty, key concerns are:

(a) High impact of a collision, grounding, explosion or fire on the business:

- Risk to life, asset, costs and environment; protecting shareholders investments; and maintaining reputation and brand.
- Strain on resources, manpower and technical capabilities.
- Achieving quick recovery.

(b) Inability to respond effectively and identify the best course of action with confidence:

- Insufficient levels of emergency preparedness and lack of experience to resolve the variety of technical issues of casualties.
- Unavailability of suitable software and ability to model scenarios correctly and accurately.

(c) Loss of reputation.

Consider the following real example. A container ship after leaving and not far from a busy port touches bottom. Double bottom tanks flood quickly and water levels are rising slowly in three holds. Will the vessel sink? Could she capsize? She is put aground and the flooding completes. Now she is well aground forward, but with each tide cycle she is settling slowly into the seabed. This will be a challenging salvage case. What additional support can be provided: to the stricken vessel; to her recovery; saving cargo and to restoring her back into to service? The response to these challenges is explored later in the paper.

It will be clear that broadly the same technical challenges apply to all ship types.

3. MEETING THE OPERATORS' NEEDS

ERS capabilities were largely defined 30 years ago. Lloyd's Register was approached by operators to support them in managing the risks after recognising the possibilities offered by the digital computer to assess damage residual strength and stability, evaluate groundings and cargo loss.

The 24/7 ERS concept was thus developed. It would be based on pre-prepared computer models. And the need for accurate computer models soon became apparent due to the sensitivity of results and compounding of errors arising in modelling complex scenarios.

Lloyd's Register's SERS was operational in mid 1980s and growing. Some years later, following the EXXON VALDEZ disaster, the USA Oil Pollution Act (OPA) 1990 [1] caught up with industry best practice requiring that, "... tankers and offshore oil barges shall ensure that their vessels have prearranged, prompt access to computerised, shore-based damage stability and residual structural strength calculation programs." This validated the original vision.

ERS providers offer the following that would generally not be available in-house:

- Shared experience of a subscription service.
- Prompt response.
- Independent support.
- Expert support.
- Experienced teams.

The following sections illustrate the challenges, risks, solutions and benefits of ERS for various ship types and scenarios

4. BULK CARRIER

4.1 CASE 1

The vessel, loaded with coal, ran heavily aground in fair weather on an offshore reef en-route to a power station terminal, (Similar to Figure 3). Damage to the bottom and flooding escalated greatly with deteriorating seas.

Figure 3 Bulk carrier aground

Faced with the possible loss of the ship and cargo, the client needed to understand the immediate and worsening strength situation, have the ship refloated in ever-changing conditions, and obtain permission to enter port and discharge. The ship would need temporary and full repair voyages.

Challenges, risks and solutions of this scenario are outlined in Table 1.

Stage	Challenge	Risks	Solution and benefits
Initial grounding.	Would swell and tide refloat damaged vessel? Strength aground. Stability and strength if self-refloats. How can vessel be refloated safely?	Delays. Increased damage. Break-up. Stability inadequate on refloat.	Calculate grounding forces. Calculate damage residual strength. Calculate free-float damage strength and stability. Calculate minimum offload plan.
Deterioration of grounding with swell, additional flooding and ship moving.	Strength aground. Stability and strength if self-refloats. Ballast down to avoid further damage? How can vessel be refloated?	Delays. Increased damage. Break-up. Stability inadequate.	Calculate grounding forces. Calculate damage residual strength. Calculate free-float damage strength and stability. Calculate minimum offload plan. Revise the above with regular updates in case of vessel movement, taking into account tide.
After refloat further slow flooding occurs after patches fail or embedded coral plugs work free. Deck is awash forward.	Could flooding threaten hatch cover integrity. Improve freeboard? Will structure resist air pressurisation applied to flooded tanks?	Stability inadequate. Sinking. Break-up.	Calculate final flooding condition. Investigate counter-ballasting. Pressurisation – local strength assessment by Classification specialists.
Get vessel alongside for discharge.	Port authority concern about entry of damaged vessel. Draft restriction.	Discharge outside port would increase costs and delay.	Confirmation of damaged strength and stability for transit to the berth and during the discharge operation supported the request for permission to enter terminal. Escalation of flooding or pressurisation failure cases were also considered, together with effectiveness of contingency action.
Return to service.	Transit to suitable repair yard(s).	Delay. Sinking.	Supported case for temporary and final repair voyages by calculation of strength and stability, including for escalation and contingency action.
Salvage plan support.	Assisted salvors at all stages. Assurance to operator. Assurance to authorities.	Permissions, delay.	Early calculations were made at Salvors request. Validation of salvor's calculations when their model was ready was carried out at their request. Validation of salvage plan at request of port/terminal operator. Validation of salvage plan at request of operator.

Table 1 Bulk carrier grounding – challenges, risks, solutions and benefits

4.2 CASE 2

A 40 000 DWT vessel loaded with steel products was struck by another ship then was run aground due to concern over the flooding. She suffered extensive damage to side of vessel in way of flooded No1 hold (Figure 4). The vessel was obstructing a busy channel and authorities required that she be removed as soon as possible, otherwise they might have the vessel destroyed

Figure 4 Bulk carrier hold damage

Challenges:
Ensure that the vessel was not destroyed and have a safe offload plan optimised as soon as possible to the satisfaction of the Authorities.

Risks:
Significant possibility of complete loss of hull, machinery, cargo and post-incident litigation.

Solution and benefits:
A minimum offload plan was calculated that reduced cargo handling time and costs. The possibility of further damage due to sloshing in the flooded hold was identified and then eliminated by further analysis.

4.3 CASE 3

A bulk carrier grounded, loaded with coal (Figure 5). The fore peak and two double bottom water ballast tanks were open to sea. Offload was not possible. She was refloated by transfer of coal cargo aft with assistance of cranes on another vessel. Calculations were made to support temporary and final repair voyages.

Figure 5 Bulk carrier aground

Challenges:
Refloating calculations of high accuracy as scope was marginal and assessment of strength and stability.

Risks:
Delays.

Solution and benefits:
Independent assurance of damage strength and stability. Calculation of refloat, short and longer repair voyages. Helped to speed up authority permissions for refloating of vessel. She was hard aground for three weeks with damage worsening in storms - early refloating reduced the likelihood of further damage.

5. CONTAINER SHIP

5.1 CASE 1

This case is the continuation of the casualty introduced in Section 2. The vessel had struck an underwater obstruction near a large busy port. Various double bottom tanks were flooded, and three holds were filling slowly (Figure 6). It was quickly identified that in this loading condition there was potential to capsize or sink. She was beached, with her bow sinking slowly into the seabed with the tide cycles and re-floating was urgently required. This was achieved over the following weeks by offloading; patching, and de-watering. The typhoon season increased the urgency.

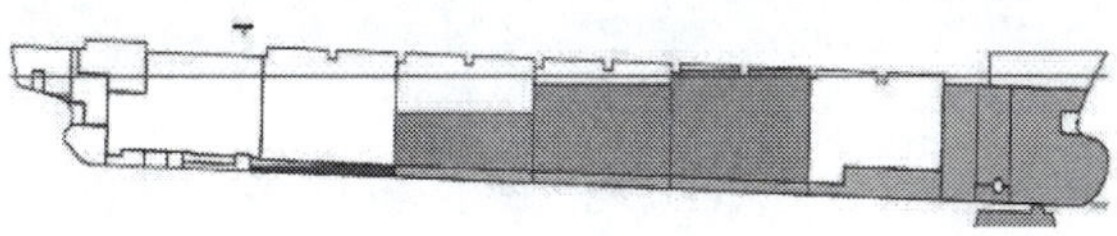

Figure 6 Flooding modelling – partial and equilibrium

Challenges:

- Rapidly assess if the ship would capsize or sink; hence the need to abandon her; and whether beaching was necessary or even achievable.
- Assist salvor who was unable to produce an accurate model while waiting on hull form data from builder.
- Assess if ship could break up.
- Determine if she can be refloated.
- Optimise cargo offload.
- Satisfy port, national authorities, insurance and other interests.

Risks:
Crew, vessel, pollution and entire cargo were threatened.

Solution and benefits:

- Stability, grounding and offloading calculations were made immediately confirming proposed courses of action.

- Initial calculations were made at salvor's request, while their models were constructed – accelerating salvor's initial assessments.
- Validation of salvor's models at their request.
- Validation of salvor's calculations at request of authorities and operator, providing assurance to them and associated interests; and assisting to obtain permissions to proceed.
- Application of experience to evaluate and optimise complex offloading/refloating sequences in the most efficient way.

5.2 CASE 2

The vessel entered a busy estuary and ran aground just after high water around midships across a sand bar. Just before low water after tide fall of 5m, she broke her back. (Figure 7) No counter-ballasting measures could have enabled immediate refloat or have prevented the failure at low water.

Figure 7 Vessel aground at low water

The vessel was refloated at high water despite the sinkage due to flooding with the assistance of an additional depth of water and aided by upward deflection of the hull girder at the grounding/failure location.

Needs:
- Refloating and reaching safe location.
- Offload and forwarding of cargo with least delay.
- Remove obstruction to the waterway.
- Manage a high-profile case under public and authority scrutiny.
- Return to service.

Vessel damage:
Hull failure corresponded to a hinge forming near the deck at the mid-hold location. The hull straightened a little due to the rise in tide from low water as the ends of the vessel gained more buoyancy. When afloat the keel line was deflected by over 2m at midships. The bottom had collapsed, and the containers lower in the hold were jammed in place as the distance between the transverse bulkheads had reduced. Further, the containers were pushing on the watertight bulkheads that were already carrying the flooding loads. The deck had stretched such that the hatch coaming had fractured in places and lengthened the coamings so much that that daylight could be seen at the hatch ends beyond the hatch cover (that retained its original length). Tanks in way of damage were ruptured.
The vessel was in a delicate and weak state and required to be treated carefully.

Risks:
Vessel break-up, significant cargo loss, delay, blockage of waterway, pollution and reputational harm.

Solution and benefits:
- Grounded condition; rapid modelling of strength and stability. Evaluation of re-floating. Consideration was given to survivability of vessel parts if she broke in two (Figure 8).
- Computer software allowed modelling of the deformed hull. Container loading tools allowed rapid development of offload plans.
- An offloading plan was calculated to release the vessel from obstructing the channel where she was confined because the sinkage due to flooding and hull deflection had increased the forward and aft draughts. The vessel was in a delicate state and the approach taken was to minimise changes to the loads, in particular stillwater bending moment, at the point of failure. Hence she could be moved from the channel to a berth in minimum time.
- The offloading plan for alongside hastened release of the cargo.
- Repair voyage, docking and repair-assistance calculations were also made. Repair included shortened the vessel to allow her to continue trading while a new hold section was built in Far East and fitted a few months later. This reduced downtime and repair costs.
- Support to salvors, making preliminary calculations on their behalf while their model was created; validating their models and calculations for themselves; authorities and operators and associated interested parties accelerated progress.

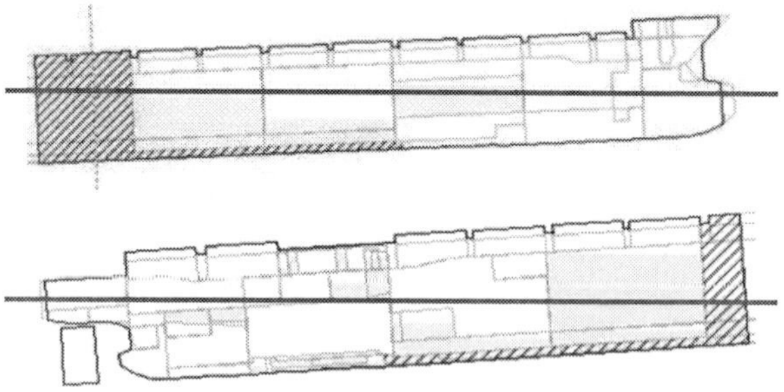

Figure 8 Vessel broken in two – fore and aft parts

6. CAR CARRIER AND RO-RO

6.1 CASE 1

The vessel had suffered loss of stability and heeled during a turn aggravated by cargo shift, then going aground lying with 45 deg. angle of heel (Figure 9).

Figure 9 Aground on sandbank

Challenges:
To help refloat and stabilise the listing vessel and salvage its cargo of 1,000-plus vehicles and construction equipment, the salvage company needed the vessel's hull form data to prepare a computer model to make calculations. Their initial calculations were made using a model of a similar ship. The hull data is protected by the builder's intellectual property rights and, realising it would take at least five days or more to obtain this vital information from the vessel's Japanese shipbuilder it was possible to provide an alternative solution to accelerate the response using the SERS model of the vessel, which had been enrolled previously.

Risks:
Inaccurate predictions could result in further damage and unpredictable behaviour during and after refloating.

Solutions and benefits:
The ERS model was used to validate the salvor's computer model and their calculations/scenarios. One of salvor's naval architects, accompanied by the special adviser for SOSREP (the UK Secretary of State for Transport's Representative) worked with the response team in the ERS offices.

The vessel was ultimately stabilised, and thousands of tonnes of clean floodwater were pumped overboard, while the mix of oil and water were pumped into one of the vessel's empty fuel oil tanks to reduce the angle of list once the vessel refloated on the high tide.

6.2 CASE 2

The car carrier suffered fire in upper decks on route to Southampton. CO_2 flooding was used to extinguish the fire, however, significant heat damage was caused to the structure, deforming and buckling several decks in the vicinity of the fire. She was allowed alongside.

Challenges:

- Offloading vehicles, including those intended for this destination and any others that had to be removed immediately due to damage.
- Onward voyage for further deliveries and/or vessel repair.
- Assessment of strength in support of above.
- Reassure authorities of proposals.

Risks:
Vessel break-up, delay.

Solutions and benefits:

- Calculation of damage residual strength.
- Evaluation of strength during vehicle offloading.
- Development of loading conditions for ongoing voyage, for further deliveries, docking and repair.

7. GENERAL CARGO

7.1 CASE 1

A 2-hold cargo ship of 8000 DWT carrying logs was involved in a collision at Shanghai anchorage flooding No.2 hold. Initially, the Engine Room was slowly flooding, but after a couple of hours the flooding was stopped and the Engine Room was pumped out. The vessel heeled 5 deg. Caissons were fitted to stabilise her during offload (Figures 10 and 11).

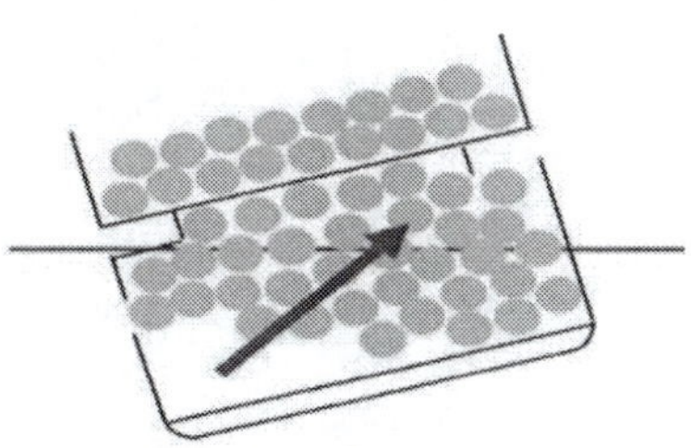

Figure 10 Cargo shift – logs floating

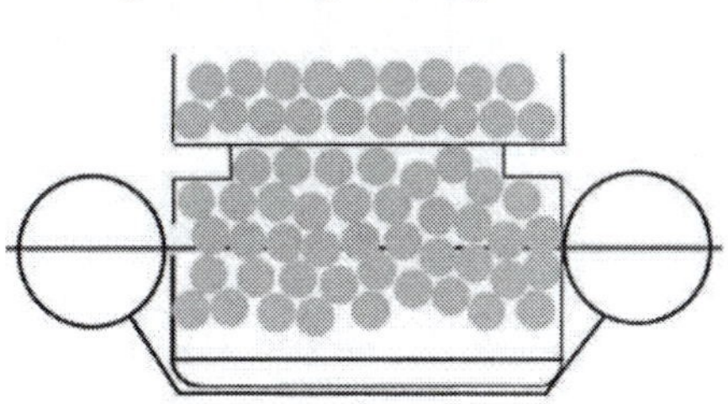

Figure 11 Caissons provided stability during offload

Needs:

- Rapid assessment of the consequences of flooding on the single-compartment standard vessel.
- Evaluation of salvor's proposals to support and move the vessel up-river using caissons.
- Evaluation of proposed offload sequence.

Risks:
Loss of stability and capsize in a major harbour, safety of life, pollution, cargo damage, delay etc.

Solutions and benefits:
- Engine Room flooding at equilibrium was shown to cause significant loss of stability; however, it was halted early, fortunately. Rapid assessment supported client decision-making
- Assessment of flooding condition. This was complicated by the floating of cargo in the damaged hold. This caused an unexpected and significant angle of heel, whereas the vessel, which was initially upright, had symmetric loading flooding and positive GM and should therefore have been upright. There was possibility of further shift and caissons were attached to stabilise the vessel during offload. Estimates of possible heeling forces of the buoyant logs supported the reasoning for the heel.
- A proposal to release the caissons for other service part way through offload of the whole cargo was evaluated and the implications of possible cargo shift were shown still to be significant. The caissons were therefore retained. Follow-up use of ERS model allowed rapid and accurate assessment of proposals to complete offloading and support proposals for the repair voyage.

7.2 CASE 2

A multi-purpose cargo/container ship (Figure 12 shows a similar type) grounded outside of a Middle East port at low water. She was unable to free herself on the high water, even with the use of available harbour tugs.

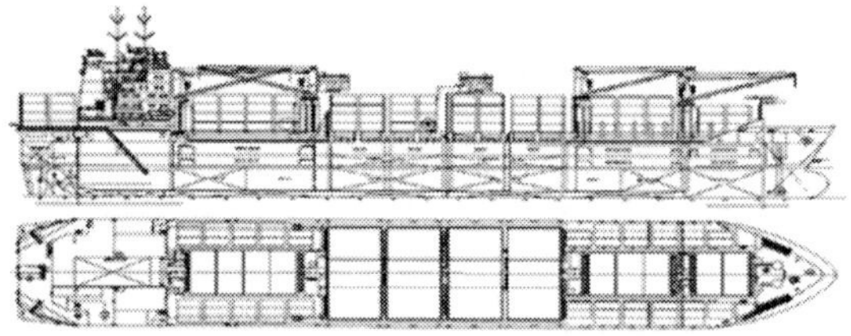

Figure 12 Typical general cargo/container ship

Needs:
Refloating.

Risks:
Incorrect action could cause damage, delay.

Solutions and benefits:
- The pre-prepared and accurate computer model allowed rapid assessment of the grounding, for which refloating without offloading was marginal.
- Although the vessel had a significant quantity of ballast available to lighten the ship, it was necessary to carry out a complex series of operations, e.g. filling some tanks to stabilise the vessel before emptying others in order to control free surface etc. in case of floating-free during preparations for final de-ballasting to float free with parallel rise etc.
- Refloating was carried out quickly without outside assistance or offloading, on the following high water, thereby minimising costs and delay.

8. PASSENGER SHIP AND FERRY

Many large passenger cruise liners are already enrolled with ERS providers and actively participate in emergency response drills and exercises. Legislation is catching up with the industry lead through requirements for decision support to masters of passenger ships and Safe Return to Port for new and existing vessels.

8.1 CASE 1

The passenger/cruise liner was 200 miles NW of New Zealand, heading for Auckland ahead of a storm, however, she had to turn into the weather and experienced very heavy rolling up to 30 degrees. This behaviour was quite unexpected and caused concern about the condition of the vessel, possibly because of unknown damage and flooding. Turning the vessel took some time, meanwhile a precautionary call was made to SERS. The rolling reduced when a heading into the seas was obtained. See [2].

Needs:
Rapid assessment of stability and investigation of possible causes of loss of stability.

Risks:
Implications of Loss of stability

Solutions and benefits:
Immediate assessment of stability followed by investigation of potential undetected flooding scenarios, e.g. of void spaces that were not fitted with detectors, and where damage control teams had not yet made access or where detection may have failed, and that might have resulted in significant loss of stability. None were found capable of significant de-stabilisation. The benefit was rapid assurance as to the good stability of the vessel.

8.2 CASE 2

A small cruise ship ran aground in the Canadian Arctic, Northwest Passage, in late summer. She struck a reef and her forward two thirds were aground with her aft part overhanging deep water (Figure 13). Most double bottom tanks in the forward part were open to sea and the loss of buoyance placed the vessel heavily aground. Passengers and non-essential crew were evacuated, and a salvage plan was commenced to refloat the vessel and take her for short voyage for temporary repair and then onwards for final repair.

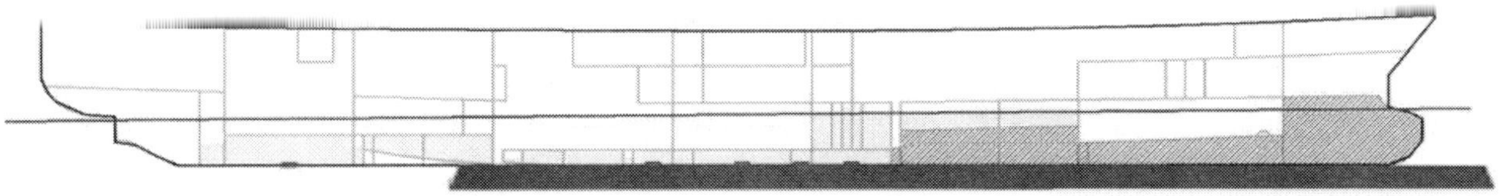

Figure 13 Cruise ship grounding modelling.

The vessel was not enrolled in SERS, but due to the sensitive location, it was requested by the authorities that she was modelled, in order to provide them with reassurance by verifying the calculations made by the salvors.

Benefits of pre-enrolment in SERS when operating a passenger vessel in remote and potentially hostile locations are not considered here.

Needs:
Client required support to validate recovery and repair of the vessels to satisfaction of local authorities.

Risks:
Further damage, break-up, sinking or capsize of vessel.

Solutions and benefits:
Independent calculations were made of damage residual strength and stability at all stages from refloating, through temporary and to final repair voyages. The benefits were reassurance about and permissions given to proceed with the operations and recover the vessel.

9. YACHT

Over fifty motor yachts are enrolled in SERS that range in length from 40m to 150m. Enrolment has been requested by vessel operators working to high standards of commercial good practice. Casualty experience is mainly of grounding. In one serious case a large yacht went aground at speed on an uncharted outcrop causing significant flooding and structural damage. Assessments of grounding, stability afloat, and support to short and long repair voyages were carried out.

The benefits of ERS have also been shown by experience of a many exercises, which reveal interesting features of the behaviour of the damaged vessels and support the understanding of the crew of this.

10. LPG CARRIER

The 30 000 DWT vessel, Type 2G with independent tanks, partially loaded with propane and butane was driven aground from her mooring by a tropical storm in the Caribbean (Figure 14). Double bottom tanks were damaged as she was pushed along the shore by the seas, eventually halting and moving apparently about a grounding point under the aft most cargo tank, port side.

Figure 14 LPG carrier aground

Challenges:

- Rapid grounding assessment.
- Carry out calculations immediately for the salvors while their model was developed.
- Assess if ship could break up.
- Determine if she could be refloated.
- Optimise de-ballasting for refloating.
- Develop offload sequence to discharge cargo.
- Develop offload sequence to allow inspection.
- Develop long repair voyage condition.
- Reassure terminal; local and national authorities; insurance and other interests.

Risks:
Break-up, pollution and delay.

Solutions and benefits:

- Grounding calculations were made for the changing situation as the vessel was moved by the seas along the shoreline.
- Grounding calculations considered worst case configurations (two-point at ends; or midships) where bottom configuration was not known.
- Effects of ballasting down to prevent further damage were considered.
- Calculations were made for the salvors who had been helicoptered onto the vessel within hours, meanwhile their computer models were prepared.
- Refloating calculations were made to free the vessel initially. The conditions had to be reassessed and revised as she grounded again a number of times while being manoeuvred out of the reef system.
- Cargo offload required careful consideration due to the major loss of strength.
- During damage inspection and survey tanks had to be emptied and a sequence was developed to minimise pumping operations. Differential

pressures on potentially weakened boundaries were minimised.

- Confirmation of short voyage docking conditions.
- The final repair voyage required accurate assessment of damage residual strength. Weather routing was used to reach calmer Equatorial latitudes in order to cross the Atlantic to Western Africa with weather routing to Portugal.
- Repair was completed with minimum delay, by an experienced yard and at less cost.

11. DEMANDS ON SHIP OPERATOR'S EMERGENCY RESPONSE RESOURCES

In making the necessary technical assessments to support their decision-making, the operator's emergency response arrangements may be faced with substantial demands. There are a number of questions that need to be asked in preparing and responding to a ship incident and having the ability to make the right choices under pressure:

- What is the availability and ability of staff for immediate response?
- Can response be sustained for long periods?
- Have they got the right support and tools?
- Do they have a suitable model ready? If not, calculations could take 24-48 hours to start, including by salvors.
- Have they got sufficient experience in dealing with any real situation that may occur?
- Will they meet response time requirements, such as those of the US Coast Guard?
- Will they be recognised by national authorities or Flag administrations?

ERS providers address these questions with dedicated, proven arrangements and highly experienced staff that may be difficult to replicate in-house for the same cost.

A dedicated ERS Response Centre is shown in Figure 15.

Figure 15 Emergency Response Centre (SERS)

Even operators of many of the world's largest fleets enrol with ERS providers to support their own arrangements in line with best practise.

Further detail on experience of development of emergency response arrangements is provided in [4.].

12. SATISFYING USCG NON-TANK VESSEL RESPONSE TIME REQUIREMENTS

Following from the OPA-90 regulations for ERS provision for tank vessels [1], the risks associated with large amounts of bunker oil in non-tank vessels (NTV) was addressed in 2005 legislation [3], applying to all vessels over 400GT carrying oil of any kind as a fuel for propulsion.

The NTV legislation does not specifically state that access to damage residual strength and stability assessment using pre-prepared computerised models is required. However, it does state the response time requirements for these assessments:

- 3 hrs to start calculations
- 12/18 hours to complete for near/offshore areas

These assessment timescales would be difficult to meet without using pre-prepared computer models which may pose a risk of legal penalties.

13. ERS AND THE SALVOR

ERS providers are complementary to and can support salvors in various ways. Clarification below of their roles may be useful.

The ERS role is that of immediate response using pre-prepared computer models to assess damage residual strength and stability, grounding and oil outflow. As mentioned in the cases above, ERS assessment may enable a vessel to be refloated before a salvor needs be engaged.

In serious cases when a salvor has been appointed then they will assemble the necessary resources to recover the vessel to a place of safety, while minimising pollution. In general, a computer model will have to be created to calculate damage residual strength, stability and grounding. Existing models could be modified or hull form geometry obtained from the builders and it could take significant time to generate an accurate model. Meanwhile, the ERS provider's model could be used to carry out calculations on their behalf, subject to permissions of the ERS contract holder.

Further the ERS provider may validate the salvor's model and calculations to provide the benefit of reassurance to:

- Salvors, at their own request
- Authorities
- Operators and their associated interests

14. CONCLUSIONS

Casualties will happen and when they do operators will need reliable technical support. There are many benefits of ERS for non-oil tankers in being prepared for, stabilising a casualty and in the post-emergency stages of offloading /returning her quickly to service.

Demands on the operator's emergency response resources may be very high and difficult to arrange in-house. In this respect ERS providers offer the following particular benefits:

- Shared experience of a subscription service
- Prompt, reliable response with constant availability
- Quality pre-prepared models, accurate simulations
- Independent, expert support and software
- Experienced teams
- Confident, informed decision making

As illustrated by the cases above, these include rapid assessment of damage residual strength and stability, grounding and assessment of oil outflow/water ingress.

The services offer assurance to authorities to enable the progress of emergency and post-emergency operations, also reducing potential delay.

Complementary support to salvage is also available.

In addressing the title of this paper, the authors have endeavoured to illustrate the benefits of ERS to non-oil tankers and identified that the majority of ships are not enrolled. The consequent risks to life, environment, assets, and reputation of the marine industry remains. It is considered that the issue is still overlooked and may not be appreciated until a serious incident occurs.

15. REFERENCES

1. US Government Publishing Office, 'Electronic Code of Federal Regulations, Title 33, Chapter I, Sub-chapter O, Part 155, Subpart D Tank Vessel Response Plans for Oil', 2015.
2. Report on the investigation of heavy weather encountered by the cruise ship Pacific Sun 200 miles north east of North Cape, New Zealand on 30 July 2008 resulting in injuries to 77 passengers and crew Marine Accident Investigation Branch Carlton House Carlton Place Southampton United Kingdom Report No 14/2009 June 2009
3. US Government Publishing Office, 'Electronic Code of Federal Regulations, Title 33, Chapter I, Sub-chapter O, Part 155, Subpart J Nontank Vessel Response Plans', 2015.
4. Lessons learnt by SERS from 40,000 years of client support: The emergency response chain. D Widger, W Gibbons, C Fisher and C Chaelis, Lloyd's Register, UK. Damaged Ship III, 25-26 March 2015, London, UK.

16. AUTHOR BIOGRAPHIES

David Prentice is a Lead Naval Architect at the Ship Emergency Response Service, Lloyd's Register, since 1999. Previous experience includes technical risk assessment, offshore structural design appraisal and trainee ship surveyor at Lloyd's Register.

David Widger is Lead Business Development, Training and Exercises and Lead Naval Architect at SERS with 7 years as a Deck Officer and 21 years at SERS. Other experience at Lloyd's Register includes field survey and structural design appraisal.

EVALUATION OF RESIDUAL SHIP STRUCTURAL INTEGRITY FOLLOWING DAMAGE USING WHOLE SHIP FINITE ELEMENT ANALYSIS

K Stone, Stone Naval Consulting, USA; **T McNatt,** MAESTRO Marine, USA; **N Danese,** NDAR, France

SUMMARY

Immediately following major damage to a ship from collision, grounding or weapons effects, the first concern is quite properly containing the fires and/or flooding and ensuring that the ship remains afloat, upright and stable. It is now routine practice to use onboard damage control and stability software to aid in these efforts. But what about the ship's residual structural integrity, including hull girder strength, and proximity to further buckling or progressive collapse? The answers to these questions are not routinely available following a casualty at sea to either the ship operators or the support organization ashore. This paper presents a method using full ship finite element analysis to conduct damaged ship residual structural evaluation, and to compute operational limits to preserve structural integrity of the damaged ship in a seaway.

1. INTRODUCTION

The immediate response by a ship's crew to damage from grounding, collision, fire, or weapons has been improved upon by mariners since man first went to sea. Naval "damage control" teams train regularly to fight fires, control flooding, restore power and return the ship to the fight if the casualty is not too great. But what happens when the damage is so severe that the ship has sustained catastrophic damage such as a major collision or significant grounding? The ship is afloat and stable but now the master and owner/naval authority must decide if they are going to return to home port or to the nearest safe harbour with significant structural damage. Is it capable of going home under its own power or must it be brought back with a heavy lift ship?

Figure 1: A damaged naval ship under tow.

The typical process currently used to determine a ship's residual strength following damage is to recalculate the ship's structural moment of inertia and section modulus at the applicable vertical stations in way of the damage. This is done by removing damaged and other no longer longitudinally effective structure from the station's section modulus calculations, e.g. plating, stiffeners and longitudinal girders. In parallel, the ship's stillwater hydrostatics and vertical bending moments are recalculated taking into account any flooding and grounding effects that came about from the grounding/collision. This simple procedure allows for the rapid determination of the longitudinal stresses in the damaged condition. For many cases of damage this simple assessment may be sufficient.

What if the ship sustained damage along the length of the hull with significant flooding and the ship could not be sufficiently repaired before it returns to its homeport under its own power or under tow? How does the owner/naval authority evaluate the ship's ability to withstand the effects of the waves and resulting motions and accelerations? What does one do if the stresses in the damaged condition start to approach allowable limits? Ships often do not fail due to the structure yielding but from buckling. Evaluation of buckling, if any, is limited to manual calculations of primary stresses only as there is no ability to calculate true 3D stresses. There is no evaluation of the effect of internal hydrostatic loads in combination with hull girder bending. And almost certainly there is no ability to evaluate the ship's residual strength against anticipated seaway loads immediately after the damage and while in transit to safe harbour. We are proposing that there is a better, more accurate method available: using existing finite element models to conduct a rapid and rigorous evaluation of the ship's residual strength and its response to anticipated seaway loads immediately following the damage event and following voyage repairs.

2. GLOBAL FINITE ELEMENT MODELS, LOADS AND ANALYSIS

Finite element analysis (FEA) has become nearly ubiquitous in ship design today and often the models are of the entire ship. The finite element mesh, i.e. the size of the elements, can range from a coarse mesh with one plate element between transverse frames and its stiffeners either smeared into the plate or lumped at the panel edges, to a finer mesh with multiple elements between frames and stiffeners modelled explicitly. Regardless of the mesh size, accurate loading of the model is often problematic for general purpose finite element codes. The following loads must be accounted for:

- Lightship mass, including discrete masses such as propulsion and major weapons systems
- Tank mass and pressure loading including the effects of accelerations
- Hydrostatics (still water and static waves)
- Hydrodynamic loads (pressures and accelerations) from waves

In addition, models of damaged ships must be able to accurately account for the damage itself, grounding, flooded compartments, hydrostatic pressures on internal spaces and the revised mass properties for determination of the ship's damaged hydrostatics and motions in waves. There are finite element codes that can handle only some aspects of the above but might require additional calculations and properties to be calculated by hydrostatics codes such as GHS™ but the MAESTRO finite element analysis and structural evaluation and optimization code can uniquely accommodate all aspects of the analysis of damaged ships, including integrated hydrodynamic loadings.

MAESTRO, which is in use by naval design authorities, shipyards and naval architects worldwide, allows for the accurate modelling of an entire ship at either coarse or fine mesh. Designed from the beginning for the analysis of ships it facilitates the modelling of stiffened plate structures, lightship and full load masses, tanks (intact and flooded), automatic calculation and application of hydrostatics and static waves (i.e. trochoidal or sinusoidal hogging and sagging waves) and fully integrated seakeeping/hydrodynamic loads from a user defined sea state spectrum using both frequency and time domain methods. Most importantly it also evaluates the entire structure against allowable stresses and failure modes such as plate buckling, stiffener tripping and ultimate strength of stiffened panels, grillages and full ship cross sections. MAESTRO will be used here to demonstrate the analysis and evaluation of a theoretical collision between a naval frigate and another ship. Figure 2 shows a typical full ship MAESTRO model of a frigate. This model is of a notional frigate designed using the NAPA Steel ship design system and then exported to MAESTRO.

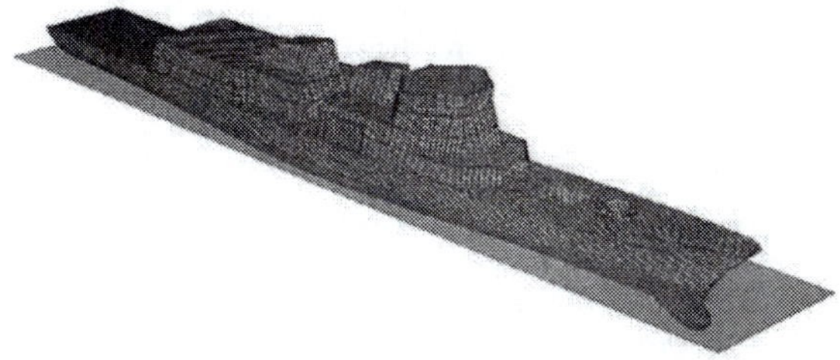

Figure 2: MAESTRO FE Model of a frigate

This model's lightship mass has been modelled by a longitudinal weight distribution. The mass at each section is applied to each node in proportion to its self-weight mass. The liquid loads in tanks are defined as per the ship's operating condition.

3. MODELING OF THE DAMAGED SHIP

3.1 MODELING STRUCTURAL DAMAGE

The damage scenario considered for this analysis is a theoretical collision between a frigate and a containership. The frigate, hit amidships in way of a watertight bulkhead, suffers significant damage below the waterline due to the containership's bulbous bow as well as damage to the main deck and superstructure. There is flooding of two adjoining main watertight compartments as well as to an inner bottom tank on the affected side. Figure 3 shows the extent of the assumed damage. All damaged structure has been removed from the model with the assumption that dished in or deformed plates adjacent to holes are no longer effective. It should be noted however that one does not have to decide what intact structure in way of the damage is effective, i.e. shadow zones, the analysis will show what structure is still load bearing.

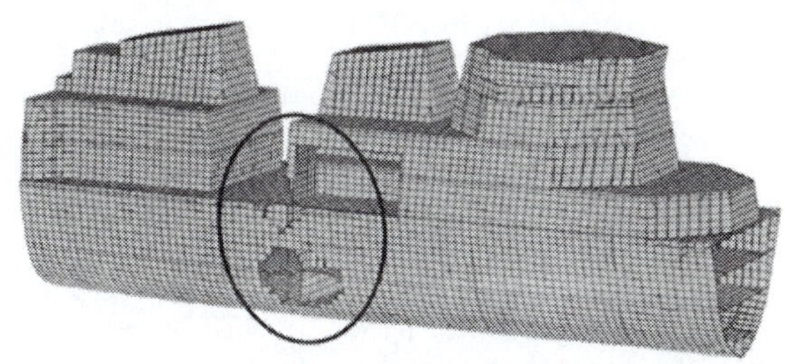

Figure 3: Damaged FE model, amidships

In this example of a collision the damage would be relatively easily observed and thus modelled accurately. If one is dealing with a grounding the determination of the damage to the hull may have to be estimated until a more complete assessment can be made.

3.2 MODELING THE DAMAGED INTACT CONDITION

In addition to the structural damage, the analysis must also consider the condition of the tanks and/or cargo at the time of the collision, the flooding of spaces as well as damage to tanks. MAESTRO allows for defining spaces and tanks as intact or flooded which thus permits the program's built-in hydrostatic balancing to reflect the additional mass as well as the hydrostatic pressures resulting from the internal flooding.

The damaged ship has a draft of 4.7m (0.7m greater than intact) with a heel of 4 degrees and is slightly bow down, a change of 0.5 degrees. As expected, the still water bending moment and shear force magnitudes and distribution are significantly different from that of the intact ship. The intact ship at full load has a maximum bending moment of 9,326 tonne-m (hogging) with negligible shear amidships and about 210 tonnes of shear at ¼ and ¾ LBP. The damaged frigate has 1,560 tonne-m of *sagging* bending moment combined with 338 tonnes of shear in way of the damage amidships. Figure 4 shows the difference in the vertical bending moments' magnitudes and distribution.

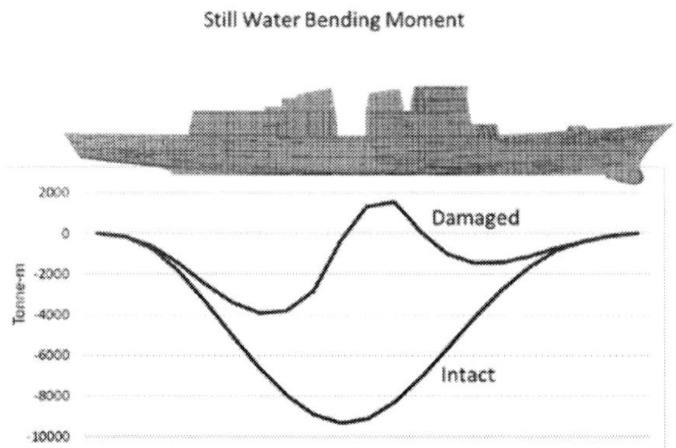

Figure 4. Intact and Damaged Still Water Bending Moments

All this information up until this point could have been obtained using a spreadsheet and a standard hydrostatics program and the stresses at the deck and keel resulting from hull girder bending calculated. But what about the loads that will result from the local sea state and the anticipated seas in the coming days as the ship returns to safe harbour either under its own power or under tow? The global stresses can thus be estimated but what about local stress concentrations, stresses arising from flooding and the margin against failure from buckling? The answers to these questions are available only with the global analysis.

4. PREDICTION OF ENVIRONMENTAL LOADS

Wave bending loads of the ship's actual and anticipated sea states must be considered. Ideally, one should evaluate the ship's motions in fully developed seas, i.e. using frequency and/or time domain hydrodynamic analyses, to calculate the resulting pressures and accelerations. At a minimum, one should consider quasi-static loads, i.e. balancing the damaged ship on waves of varying heights and lengths. These are relatively easy to model.

4.1 QUASI-STATIC WAVE LOADS

Quasi-static hull girder bending loads of anticipated waves should include the hydrostatic pressures and be fully balanced. Waves of varying height, length, phase angle (i.e. varying the position of the wave crest/trough) and angle on the bow should all be considered. Determination of the ship's draft, heel and trim on waves and the resulting hull girder bending moments are easily obtained from hydrostatics programs. Care must be taken to balance the FE model on these waves to have a fully balanced model. In MAESTRO's case, these calculations are done by the program and a fully balanced model results.

Table 1 shows the FE model of the damaged ship on a quasistatic bow quartering wave along with the resulting vertical bending moments amidships in way of the damage for a range of quasistatic waves.

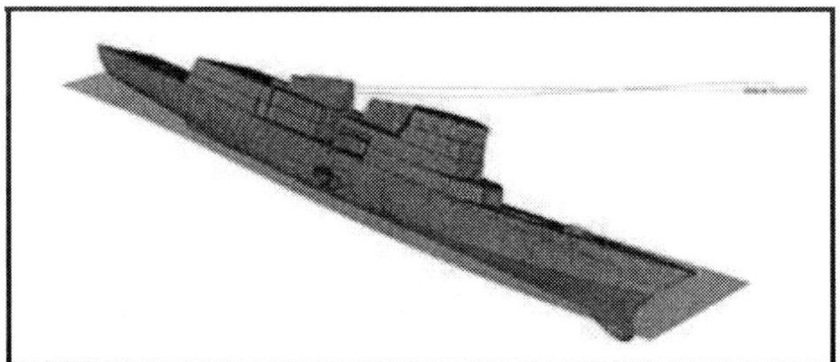

Condition	Bending Moment (tonne-m)
Stillwater, sagging	1,340
7m height, 135m L, sagging	23,331
7m height, 135m L, hogging	16,913
3m height, 135m L, sagging	10,197
3m height, 135m L, hogging	7,054
3m height, 135m L, quartering, sagging	4,055
3m height, 135m L, quartering, hogging	1,009
3m height, 67.5m L, crest amidships	1,880
3m height, 67.5m L, trough amidships	1,003

Table 1. Quasi-static wave loads, damaged ship

4.2 DYNAMIC WAVE LOADS

Calculation of hydrodynamic loads is now in the toolbox of many naval architects and design authorities however the application of those loads, the panel pressures, and rigid body accelerations, to a finite element model is still generally fraught with difficulties. General purpose seakeeping codes calculate their loads on a surface model of the hull using an internal mesh and produce plots of the hull girder bending moments and shear, the ship's accelerations and pressure distributions. It is extremely difficult to translate these loads, which are balanced, to a finite element model which has a different mesh. An integrated approach which automatically translates the loads and balances the model is needed. MAESTRO has a fully integrated potential flow-based calculation and application of hydrodynamic loads using both frequency and time domain methods for user defined sea states, wave spectra, and operating profiles. The calculation of the hydrodynamic pressures is done on the FE mesh which results in full equilibrium of the model.

The MAESTRO Wave program calculates the ship's response to a user defined wave scatter diagram, wave spectrum, and operating profile. In this example study, we calculated the response of the damaged ship, including the inertial conditions caused by flooding, in 1-3m waves (sea states 3-5) at 0 and 5 knots, in head through stern seas, with a frequency domain analysis using a North Pacific wave scatter diagram, Bretschneider wave spectrum and a simplified operating profile of equal probability speeds, headings and wave heights, for a 3-hour period. MAESTRO will calculate and determine the user chosen "dominant load parameters", i.e. worst case, for bending moments, shear, motions, etc. at each combination of heading, speed and wave height. Figure 5 shows the pressure distribution (wave only and then combined pressures) for the maximum vertical sagging bending moment in head seas at 5 knots.

Wave pressure distribution

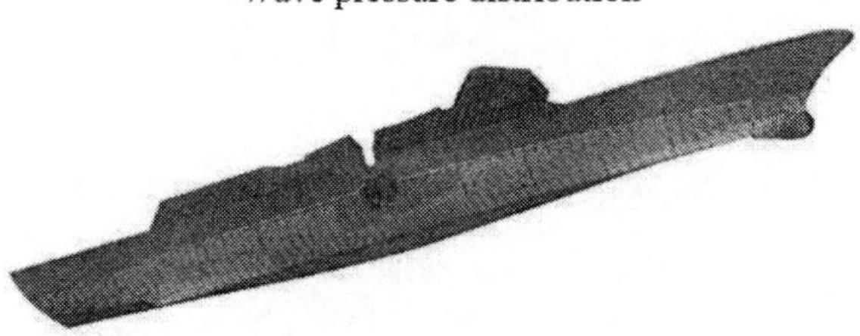

Combined (net) pressures (stillwater, wave, flooding)

Figure 5. Frequency domain pressure distribution: maximum vertical bending (sagging), head seas, 5 kts, wave period 9 secs, wave height 2.2m

In this case we considered only the vertical bending moment. Table 2 lists the maximum bending moments amidships for the damaged ship at stillwater and at 5 knots in 3m H_S waves at various headings. For comparison, the maximum sagging and hogging bending moments for the damaged ship on a quasi-static 3m trochoidal wave resulting from a 3m trochoidal design wave are 10,197 tonne-m and -7,054 tonne-m, respectively whereas the frequency domain analysis's head seas case calculated higher sagging and bending moments, 14,482 tonne-m and -12,782 tonne-m, respectively. The dynamic analyses' higher, and more accurate, bending moments will result in higher stresses and potentially more areas where the structure fails to meet its safety criteria.

Condition	Bending Moment (tonne-m)	Sagging Hogging
Stillwater	1,340	Sagging
Heads seas, 5 kts	14,482 -12,782	Sagging Hogging
Bow quartering, 5 kts	13,500 -11,800	Sagging Hogging
Beam, 5 kts	4,742 -3,041	Sagging Hogging
Stern quartering, 5 kts	9,889 -8,188	Sagging Hogging
Stern seas, 5 knots	12,209 -10,328	Sagging Hogging

Table 2. Damaged ship bending moments obtained from frequency domain analysis, 3m H_S seas.

5. ANALYSIS AND EVALUATION OF THE DAMAGED SHIP

To be truly useful, the finite element analysis of the damaged ship must include the ability to comprehensively evaluate the structure against failure modes, or limit states. These limit states must include yield-based limits as well as complex buckling of the stiffened panels, framing and girders. Our proposed method conducts these evaluations automatically. The failure criteria used here are MAESTRO's native limit states as described in Hughes [1] and Hughes/Paik [2]. Other available criteria include Paik's ALPS/ULSAP [2] as well as well-known naval and classification authority standards including ABS's High Speed Naval Craft rules, ABS/USN's Naval Vessel Rules, ABS's Steel Vessel Rules, the U.S. Navy's DDS100-4, Bureau Veritas's Naval Ships Rules, and the Common Structural Rules for Oil Tankers. The evaluation of the complete structure for all load cases takes just a few minutes vs. the many hours it takes to manually check a general purpose FEA's results against the required safety criteria.

MAESTRO uses a metric called the "adequacy parameter" to evaluate limits states. If Q is the load effect, Q_L is the limit value and γ is the factor of safety then the safety requirement is

$$\gamma Q \leq Q_L$$

and the strength ratio, R, is defined as

$$R = \gamma Q / Q_L$$

MAESTRO's adequacy parameter, $g(R)$, is

$$g(R) = (1-R)/(1+R)$$

which varies from -1 (unsatisfactory) to +1 (satisfactory) with a value of "0" indicating that the safety factor has just been satisfied. The factor of safety can vary for each limit state. In this example we relied upon the advice of a professional salvor who recommended that we use a factor of safety of 1.25 for yield-based limits and 2.0 against buckling. If we were evaluating the ship against more benign conditions it might be allowable to decrease the factors of safety, particularly against collapse. Hereinafter we will use the term "failure" to those instances where the structure does not meet the required factors of safety against yield and buckling.

It must be noted that MAESTRO is a linear elastic finite element code; although it predicts panel buckling and yield it is not a non-linear FEA code, its FEA does not take the structure through plastic failure. MAESTRO's limit state evaluation algorithms however do include non-linear aspects of the failure modes. For our purposes that is sufficient, we are evaluating the strength of the ship's damaged structure and are striving to prevent further failures. If one needs to conduct a non-linear analysis of a structure, we would suggest that a linear analysis be conducted first to inform the more complicated non-linear analysis to follow.

5.1 ANALYSIS AND EVALUATION USING QUASI-STATIC WAVE LOADS

A finite element analysis and structural evaluation was conducted on the damaged ship using quasi-static wave loads summarized in Table 1. The 7m hogging and sagging waves, equivalent to the intact ship's design hull girder loads, resulted in several failures in the main deck and shell in way of the damage. This was as expected. More interesting however were the failures resulting from the 3m waves.

As shown in Figure 6a there are no yield-based failures in the shell between the upper and lower damaged areas (negative values of adequacy parameters, shown in ochre to red). Examination of the collapse failure modes, however revealed that the shell is exceeding its factor of safety, 2, against collapse. Figure 6b shows that there are significant areas of not meeting the factors of safety in the shell in way of the damage, specifically against panel collapse, combined buckling and panel collapse, stiffener flexure [1], which are both significant failure modes. Checking the structure against yield alone would have missed this potentially critical structural weakness.

This analysis and evaluation of the damaged ship using quasi-static wave loads has proven itself to be a valuable tool when dealing with a severely damaged ship in waves. We will now turn to the analysis and evaluation of the damaged ship using the loads calculated by the frequency domain hydrodynamic analysis.

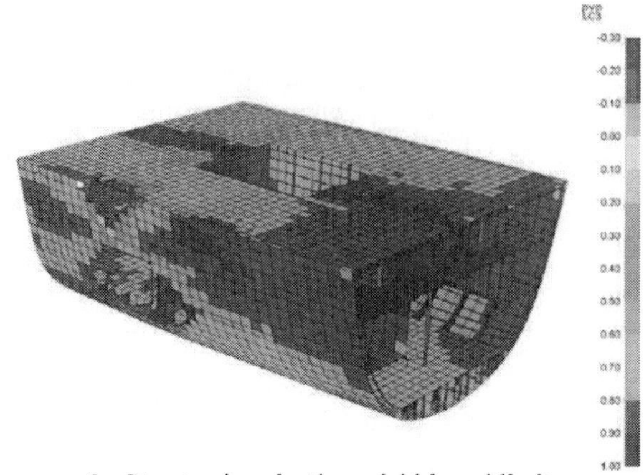

6a. Structural evaluation, yield-based limits

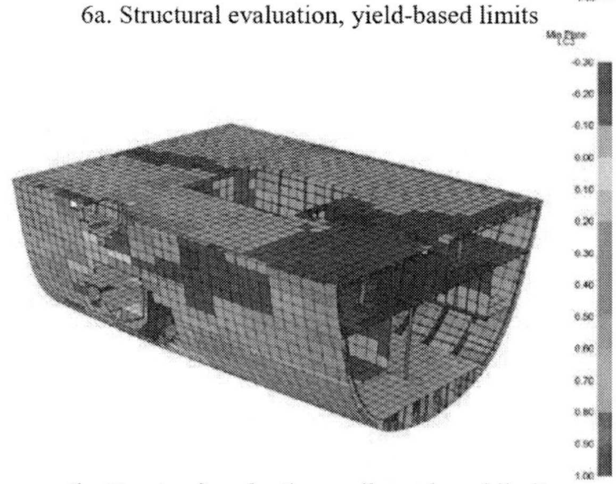

6b. Structural evaluation, collapse-based limits

Figure 6 and b. Structural evaluation, plate limit states, 3m trochoidal hogging wave

5.2 ANALYSIS AND EVALUATION USING FREQUENCY DOMAIN LOADS

A finite element analysis was carried out on the damaged ship using the waves loads from the frequency domain analysis described in 4.2 and summarized in Table 2.

Once again, the hull girder longitudinal stresses for all load cases were well within limits but the structure did not meet the desired safety factor, 2.0, against panel buckling. Whereas the quasi-static load analysis showed limited failures on a 3m sagging wave, the frequency domain analysis for 5kts in 3m H_S head seas showed more extensive failures across the main deck in way of the damage. Figure 7 compares the results for these two load cases from the quasi-static and frequency domain analyses.

The structural panels coloured ochre to red have exceeded their safety criteria against panel collapse. As can be seen the main deck in the frequency domain analysis has considerably more failures. In the example studied here, the damaged ship's response to 3m significant wave height seas at 5 knots showed that further damage would likely ensue while in transit. If this were an actual damage response the naval authorities could then evaluate alternatives such as waiting for calmer seas, alternate routing to avoid head seas or docking the ship on a heavy-lift transport for return to home port.

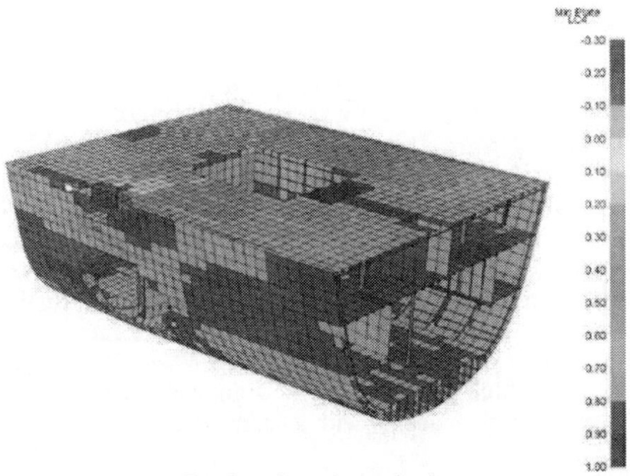

7a. Quasi-static analysis

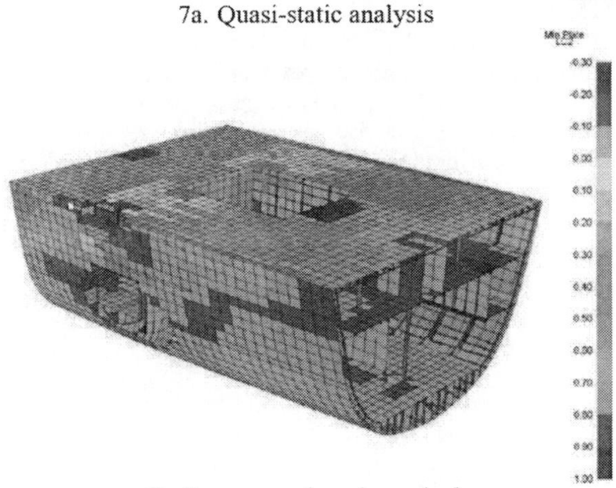

7b. Frequency domain analysis

Figure 7 a and b. Panel collapse adequacy parameters for quasi-static analysis and frequency domain analysis, sagging

5.3 EVAULATION OF HULL GIRDER ULTIMATE STRENGTH

A structural evaluation of a damaged ship's residual strength should include the calculation of the hull girder's ultimate strength, i.e. at what bending moment would the hull girder collapse?

ALS/HULL, described by Paik and Thayamballi in "Ultimate Limit State Design of Steel-Plated Structures" [3], calculates the hull girder's ultimate strength at any desired cross section. Embedded in MAESTRO, the time needed to conduct such an analysis with an existing finite element model is negligible, less than an hour.

In our example, an ultimate strength analysis was conducted of the frigate's hull girder in way of the damage. The sagging analysis, Figure 8, showed that the structure starts to undergo non-linear failure at about 55,000 tonne-m of bending with an ultimate strength bending moment of approximately 80,000 tonne-m, well above the worst case bending moments calculated by the frequency domain analysis.

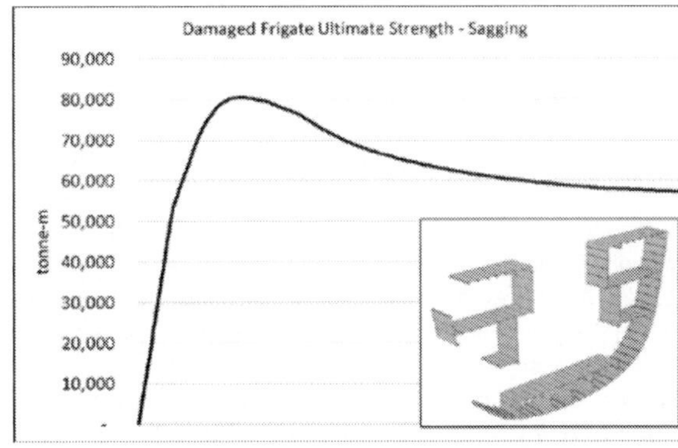

Figure 8. ALPS/HULL ultimate strength analysis of the damaged frigate, sagging

In this case the hull girder's ultimate strength is not an issue as most of the damage was to the side shell, near the neutral axis. Other damage scenarios, particularly groundings, might have a larger effect upon the hull girder's residual ultimate strength.

6. CONCLUSIONS

We have shown that there are substantial benefits to conducting a finite element analysis of a damaged ship to determine its residual strength, particularly when one is confronted with extensive damage and uncertain loads from the ship's anticipated sea states.

It is no small feat to develop a full ship finite element model. If there is no "product model" available, such as NAPA, it may take months of intensive effort to create a model using a general purpose FEA code, or several weeks using MAESTRO's ship-oriented modeler. However, many new-build projects include conducting a finite element analysis during the design process. These models can be saved and made part of the naval authority's or owner's database available for use in the event of a disaster.

Similarly, the calculation of the hydrodynamic loads is a time-consuming process even if that calculation is integrated within the finite element code. Typically, this is done during the design of the ship when one is concerned with defining the maximum loads to be encountered in the ship's life-time considering the vessel's top-level requirements, e.g. survival in sea state 7 at best heading and speed or operations in sea-state 4, all headings at 25 knots. When we are assessing the residual strength of a damaged ship we need to know the loads resulting from known or anticipated sea states on a ship with a different mass, trim and heel. We would suggest that the best time to calculate the response of a damaged ship is not while one is in midst of a crisis but in advance.

If a naval authority or ship owner has a finite element model of the ship available in the event of a collision or grounding then they should create a "damaged ship unit wave loads database" by conducting a series of hydrodynamic seakeeping and loads analyses of the ship for a range of damage scenarios, e.g. adjoining pairs of compartments flooded, at nominal speeds, say 0-10 knots, and a range of headings from heads seas to stern seas. This will allow a more rapid assessment of the damaged ship's residual strength by going directly to modelling the damage, importing the appropriate motion RAOs from the damaged scenario database, and then calculating the wave loads for the anticipated seas states.

The decision to go to a nearby port for repair or return to home port without repair must be made after a careful evaluation of the extent of the ship's damage, and of the anticipated seas during the transit. If the decision is made to return to home port without a permanent repair then one must decide whether to do so under the ship's own power, under tow, or on a heavy-lift ship. Conducting a comprehensive full ship finite element analysis as described here can go a long way to ensuring the ship's safety as well while fully informing what will almost certainly be a multi-million dollar decision.

7. ACKNOWLEDGEMENTS

The opinions in this paper are entirely those of the authors and do not reflect those of any naval authorities. We wish to thank Mr. Phil Reed of Reed Maritime Consulting LLC, and Mr. Vince Jarecki of the US Navy's Office of the Supervisor of Salvage for their advice on the practices used by salvors.

8. REFERENCES

1. HUGHES, O., 'Ship Structural Design', *The Society of Naval Architects and Marine Engineers*, 1988.

2. HUGHES, O. and PAIK, J., "Ship Structural Analysis and Design', *The Society of Naval Architects and Marine Engineers*, 2010.

3. PAIK, J. and THAYAMBALLI, A., "Ultimate Limit State Design of Steel-Plated Structures", *John Wiley & Sons, Ltd.*, 2003

9. AUTHORS BIOGRAPHIES

Kevin Stone is the owner of Stone Naval Consulting LLC and a consultant to MAESTRO Marine LLC. Prior to this he worked for over 20 years at Proteus Engineering and DRS Technologies' Advanced Marine Technology Center where he performed or directed the finite element analysis of over 50 different ships. He has a degree in Naval Architecture from the United States Naval Academy and is a Professional Engineer in Naval Architecture and Marine Engineering.

Tobin McNatt is a partner at MAESTRO Marine LLC. a firm formed in 2017 to continue the development, distribution, and technical support of the MAESTRO ship structural design and optimization technology and software. He has developed and managed technical programs, for organizations such as NAVSEA, NSWC Carderock Division, Office of Naval Research, US Special Operations Command, DARPA, and several ship classification societies. He is a graduate of the United States Naval Academy with a degree in Physics.

Nick Danese is the founder of Nick Danese Applied Research (NDAR). He has been an active user of MAESTRO for over 25 years and has participated in numerous structural analyses projects ranging from complex forensic investigations to advanced structural design. He holds a MS Naval Architecture Marine Engineering degree from the University of Michigan.

The Costa Concordia Wreck Removal. How commercially available software was customized to make mission impossible a success

P Reed, Reed Maritime LLC, USA

SUMMARY

The removal and redelivery of the M/V COSTA CONCORDIA was the biggest and most costly wreck removal project ever undertaken. Refloating the casualty was executed using external buoyancy comprised of thirty sponsons and a pair of blister tanks custom fit under the bow. A bespoke Control & Monitoring System was developed to remotely ballast 66 external tanks. The system measured tank loads, waterplane data and hull deflections, all of which were input into the Load Monitoring Program for real-time evaluation of the casualty's stability and residual structural strength. This paper discusses the challenges involved with integrating the General HydroStatics (GHS) software into the Control & Monitoring System and how it was used to plan the dewatering sequence, execute the plan and monitor the casualty during the refloating operations and subsequent tow.

NOMENCLATURE

CMS	Control and Monitoring System
GHS	General HydroStatics software
LCG	Longitudinal centre of gravity
LE	GHS Load Editor module
LMP	Load Monitoring Program
ROC	Remote Operations Centre
RTM	Real Time Monitoring
TCG	Transverse centre of gravity
VCG	Vertical centre of gravity

1. INTRODUCTION

The capsizing of the M/V COSTA CONCORDIA on the Isola del Giglio in January 2012 resulted in 32 deaths and made headlines around the world. Salvage contractors would work for the next two and a half years to remove all 45,500 tonnes of the 290-metre-long casualty in one piece and redeliver her afloat in Genoa. It would take another three years to scrap the vessel and contractors were still working in 2018 to clean up the wreck site.

The wreck removal was completed in two major steps; parbuckling and refloating. The extent of damage to the vessel and the near impossibility of establishing watertight subdivision within the wreck forced engineers to refloat the casualty using only external buoyancy. The external buoyancy was provided by 30 sponsons including twelve *tall* sponsons, four *horizontal* sponsons and fourteen *short* sponson. Two large blister tanks were fitted under the bow. The total weight of the sponsons and blister tanks was ≈18,000 tonnes.

A compressed air system was used, in lieu of pumps, to deballast the tanks. There were 66 tanks to be ballasted/deballasted which required 528 valves to be operated and 264 pressure sensors to be monitored. Unlike a typical refloating where only a half-dozen tanks are dewatered manually by the salvage team, engineers decided that it was necessary to perform this refloating from a single point that became known as the Remote Operations Centre (ROC). To do this required a bespoke Control & Monitoring System (CMS) which allowed the engineers to ballast and deballast tanks without leaving the ROC.

Each sponson had a dedicated electro-pneumatic PLC board to control each of four sea chest valves and twelve compressed air valves per sponson. The PLC boards also monitored four pressure sensors in each tank to determine the volume of water in the tank. Two pressure sensors mounted externally to the bottom of the sponsons were used to define the *measured* waterplane of the casualty and to determine the vertical position and the buoyancy of the individual starboard sponsons. All of the sensor data was collected in real-time and stored in data files on the CMS server.

A Load Monitoring Program (LMP) was developed to read the data files provided by the CMS and to calculate the residual stability, bending, shear and torque during the refloating and subsequent tow to Genoa. Output from the LMP was collected and displayed in the ROC. Third party graphs and presentations of the output data were displayed in the ROC for monitoring the condition of the wreck (Figures 2 & 3).

The CMS was designed with redundancy on generators, compressors, valves, sensors and servers. The LMP had the ability to input data collected manually from around the casualty in the event of a partial or complete failure of the CMS.

The primary purpose of the LMP was to provide real-time information to the salvage team during the operations to ensure a safe refloating. The secondary purpose of the LMP was to present the data to the owner's representatives and the Italian authorities in a manner that they could readily understand and to maintain their confidence in continuing the operations.

The LMP was developed in much the same manner as one would put together a typical loading program for an intact vessel. IACS rules for such systems were used as a basis for the LMP design. There were several algorithms that would not normally be included in typical loading program. These bespoke algorithms are discussed throughout this paper.

2. SOFTWARE & HARDWARE

The underlying software used for the LMP was General HydroStatics (GHS) from Creative Systems, Inc. GHS is a leading stability and longitudinal strength software with over 1800 individual users in nearly 40 countries. It is used by design firms, salvage companies, coast guards, navies, shipyards and universities. GHS output is accepted by all the major classification societies. Many salvage contractors use this software for planning their salvage and wreck removal projects.

GHS employs first principle calculations based directly on a 3-D geometric model of the vessel and the loads imposed on it. Output includes but is not limited to hydrostatic properties, stability, shear, bending moments and torsion. GHS is an excellent tool for calculating the damage stability and residual structural strength of a casualty for any given loaded condition and waterplane.

The GHS Load Editor (LE) module was used as the graphical user interface throughout the operations (Figure 1). LE provided a visualization of the current situation and access to a bespoke menu system that can be seen along the left-hand side of Figure 1. Data transfer and processing was performed in the background.

The bespoke menu was written specifically for this project using the programming tools (e.g. variables, macros, functions etc.) available within GHS. The menu was comprised of 17 separate files and over 7,000 lines of code. The menu ran on top of the underlying GHS software and was accessible within LE. The menu was the interface between the CMS and the LMP for data import, processing and export.

The LMP was available at two separate work stations in the ROC during the refloating, towing and redelivery (Figure 2). Both work stations were setup for two modes of operation:

- The *Prediction Mode* required the user to operate the program manually (e.g. input the sponsons tank loads) or with run files. Pre-planning the stage-by-stage dewatering of the wreck during refloating was done using the LMP in *Prediction Mode.*

- The *Real Time Monitoring (RTM) Mode* was a semi-automatic mode. In this mode, tank loads, external waterplane measurements and other input data could be imported directly from the CMS at user-defined intervals. In practice, 30 second intervals were more than adequate.

During the actual refloating operations, both LMP computers were able to operate independently in either mode as desired. Both work stations were run in *RTM Mode* unless the *Prediction Mode* was required to explore other de-watering sequences or "what-if?" scenarios during the operation.

When both computers were in the RTM Mode, one computer was designated the *Master* computer. The *Master* computer could receive input from, and return output to, the CMS. The second computer was the *Slave* computer. The Slave computer could receive input from the CMS but could not return output. The second computer provided redundancy and could easily become the *Master* computer should the first computer fail.

3. THE GEOMETRY FILE

The 3-D model of the vessel, otherwise known as the geometry file, defined the buoyant volume of the vessel (Figure 4). The buoyant volume included the hull up to deck 9, the sponsons and the blister tanks. GHS calculated the displacement and centre of buoyancy of the model at any waterplane. The customized menu allowed the starboard sponsons S3-S15 to be moved individually to simulate their vertical movements relative to the hull during the operations. Over 300 internal compartments were defined in the geometry file.

All of the compartments below deck 0 were modelled as *flooded* tanks. Their underwater volume and waterplane, if applicable, were subtracted from the buoyant volume and total waterplane area respectively (i.e. "lost buoyancy method").

Compartments between deck 0 and deck 8 were modelled by deck and fire zone except for the A/C machinery spaces, the engine room casing and the theatre. All of these compartments were modelled as *spilling* tanks to simulate "*free water*" on deck. *Critical points* were used to define sill heights in these compartments. Sill heights were set to 30mm but could be changed by the user to simulate more or less free water.

If the *critical point* for a spilling tank was below the waterline then the tank would behave the same as a *flooded* tank. If the *critical point* was above the waterline the tank would behave as an *intact* tank and the wedge of water trapped in the tank below the *critical point* would remain in the compartment as "*free water*". If the *critical point* was above the waterline and the water level in the tank exceeded the sill height, the water would spill from the tank.

All sponson and blister tanks were modelled as *intact* tanks. These tanks were loaded manually in *Prediction Mode* and automatically in *RTM* Mode with data provided by the CMS.

The port side sponsons P3-P13 were welded to the hull prior to parbuckling. The starboard sponsons S3-S15 were connected to the hull after parbuckling with a sliding connection incorporating 135mm chain and strand jacks. These sponsons could be raised and lowered relative to the hull. Several of the starboard side sponsons could not be lowered to their final design position until after refloating due to obstructions. Lowering the sponsons to their design location was necessary to affect the maximum buoyancy of the sponsons, all of which was necessary to achieve the contractual redelivery drafts. It also helped to facilitate the refloating operations by maintaining the symmetry between the port and starboard sponsons.

All starboard sponsons were fitted with two external draft sensors. The draft sensors were used to measure the vertical position of the starboard sponsons relative to the hull. The measured position of each sponson was then compared to the current position used in the geometry file. The LMP menu system provided the user a means of altering the position of these starboard sponsons (i.e. modifying the geometry file from within the LE module) while maintaining the current loading condition. Sponson related data such as its centre of gravity, its reference points and draft sensor locations was updated automatically.

If the measured position of a sponson differed from its current position in the geometry file by more than a user-defined allowable deviation, audible and visual warnings would occur. The sponson could then be moved up or down from within LE to match the positions derived from the draft sensors. If both draft sensors on a particular sponson failed, the sponson position could be evaluated by measuring the vertical distance from a known location on the hull down to a known position on the sponson. The data could then be entered manually into the LMP with the available tools.

4. LOADS

Loads on the casualty included fixed weights, both point loads and distributed loads, as well as tank loads, ground reaction and potential environmental loads (e.g. wind and waves).

4.1 FIXED WEIGHTS

Defining the *lightship* weight began with the distributed lightship weight curve. It would eventually include the weight of the sponsons, blister tanks, salvage equipment and all other weights added, removed or relocated. This included estimates of weight and centre of gravity for items lost on decks located in the "splash zone" after parbuckling. Several of these decks were completely stripped by the sea of all outfitting, including furniture, carpeting, non-structural bulkheads, toilets, showers etc. Keeping track of the true *lightship* weight was a continuous and exhaustive exercise. By the time of the refloating the net *lightship* weight had increased from ≈45,000 tons to ≈65,000 tons. Such a weight increase is not normally associated with a refloating.

4.2 TANK LOADS

Intact tank loads were computed automatically based on the following:

- The weight of seawater in a tank was determined by the volume and the specific gravity of the contents in the tank.

- Tank loads were allowed to shift with trim and heel instead of relying on a free-surface moment simulation of the shift of weight and COG. The COG of the tank load was therefore determined by the volume in the tank, the shape of the tank and the trim and heel of the vessel.

- Tank loads were distributed over the length of the tank accordingly.

4.3 GROUND POINTS

More than 50 *ground points* were defined to simulate the points of contact between the hull and anything that would restrict the vertical movement of the vessel including the seabed, platforms and grout bags (Figure 5). A first principle based "added buoyancy" and "effective waterplane" were calculated for each ground point (i.e. "added buoyancy method") and included in the total buoyancy and waterplane for the casualty. These values would be updated with each change in the loaded condition of the casualty or her waterplane.

4.4 VARIABLE LOADS

There were several identified loads that were variable and difficult to estimate:

- *Free water* was defined as water that was trapped during refloating on decks that were immerging from the water and was free to move around within its own compartment. Free water was estimated using *spilling* tanks as previously discussed.

- *Absorbed Water* was that water that had been absorbed into the mattresses, carpet, insulation etc. and was retained (i.e. not free to move) as the casualty emerged from the water. An algorithm was available in *Prediction Mode* to estimate the amount of *absorbed water* for a given waterline. The user had the option to apply a *High Contingency* of *absorbed water*, a *Low Contingency* of *absorbed water* or none at all. Some contingency for *absorbed water* was generally included when working in the *Prediction Mode*. The user could turn off the *absorbed water* algorithm in *RTM Mode* as the *Correction Factor* (*CFT*) described below would include the actual *absorbed water*.

- It was not known how much trapped air might be contained within the vessel. All tanks that were below the waterline were considered to be 100% flooded with some residual buoyancy retained depending on the permeability of the compartment. Trapped air would make the wreck appear to be lighter than it really was and could have a significant effect on heel and trim. It was very difficult to predict how much trapped air existed in what compartment. Any trapped air would have an effect on the *CFT*.

4.5 CORRECTION FACTOR (CFT)

Given the buoyant volume and loads discussed above, GHS solved for an equilibrium waterplane such that the summation of all forces and moments acting on the vessel were zero. This was known equilibrium and provided a *calculated* waterplane.

When in *RTM Mode*, the CMS provided data to the LMP from 44 draft sensors, 10 gyros and 5 GPS units. This data was used to define the *measured* waterplane.

As the casualty was dewatered and the draft was reduced, it was possible for the "*lightship*" weight and centre of gravity of the casualty to change. Any *absorbed water* would increase the weight of the casualty. Over time the absorbed water could drain from the carpet, mattresses, wallboard etc. and become *free water*. Free water would drain overboard over time and likewise decrease the weight of the casualty. The amount of *absorbed* and *free water*, along with other uncertainties (e.g. lightship weight estimates, trapped air etc.) were almost certain to result in a difference between the *calculated* waterplane and the *measured* waterplane. The LMP repeatedly compared the *calculated* waterplane to the *measured* waterplane and, in turn, calculated a weight, LCG and TCG to bring the casualty into equilibrium at the *measured* waterplane. The VCG of this weight was set to the current VCG of the vessel. This calculated weight was known as the *Correction Factor* and was shown as "*CFT*" in the list of fixed weights. The *CFT* could be positive or negative and was distributed over the length of the vessel for the purpose of shear and bending calculations.

The interpretation of the magnitude and centre of gravity of the *CFT* provided useful insights into the unknown forces and moments acting on the vessel. For example;

- If the magnitude of the *CFT* was small and the TCG well off the starboard side of the vessel then it could be assumed that the weight estimates were correct but the TCG has been underestimated.

- If the centre of gravity deviation was small and the weight was large then we could assume an evenly distributed weight such as absorbed water.

- A sudden increase in the magnitude of the *CFT* and a corresponding movement of the centre of gravity towards the bow may indicate trapped water being retained in a large compartment forward that was just now coming out of the water.

Changes in the *CFT* were tracked over time to ensure that there were no large unknown errors in the assumed weight and buoyancy of the casualty. Just after coming afloat the *CFT* was over 3000 tonnes. The *CFT* decreased to ≈1500 tonnes several days later at the start of the tow. The *CFT* decreased to ≈800 tonnes upon arrival in Genoa and accounted for less than 1% of the total weight of the casualty upon redelivery. The real-time plot of the *CFT* over time in the ROC greatly increased the confidence in the LMP.

5. STABILITY EVALUATION

The stability of the casualty was evaluated in two ways:

The LMP provided a real-time calculation of the transverse metacentric height (GMT). GMT was shown in the condition summary at the top of the Load Editor screen and could be plotted over time in the ROC.

The LMP also provided a Righting Arm Curve (GZ Curve) for any equilibrium condition. There were two options. The first option was to run the righting arm to the side of the initial list of the vessel and to evaluate the curve against the modified unmanned barge stability criterion (IMO A749(18) Ch. 4.7) as agreed by the authorities. The second option was to run the righting arm curve from 60° port to 60° starboard to see a wide range of stability for the wreck in both directions of heel.

6. RESIDUAL STRENGTH EVALUATION

The LMP calculated the shear, bending moments and torque for any equilibrium condition. The results could be shown as actual values or as a percentage of the permissible values and could be plotted over time in the ROC.

Very detailed surveys of the damage to the casualty were performed prior to the refloating. Based on these surveys, the authorities evaluated the residual strength of the casualty and approved permissible shear and bending moment curves for the refloating and tow to Genoa.

Permissible curves for torsion were not provided. The arrangement and capacity of the sponsons provided an opportunity to impose large torque loads on the casualty.

For this reason, engineers monitored these loads and minimized them to the extent possible.

7. DATA PROCESSING

Automatic data processing was only available when the LMP was in *RTM Mode*. In this mode, the CMS provided real-time data for the LMP to read and process. Results of the data processing were then sent back to the CMS for display in bespoke third-party software.

The data exchange between the CMS and LMP occurred at user defined intervals known as *idle time*. The *idle time* could be adjusted by the user depending on the nature of the current operation. Reading input files, processing the data and writing output files was referred to as "*idling*" or as an "*idle*".

The CMS would collect data and continuously overwrite a data input file for the LMP. This file was stored on the CMS servers and included:

- draft sensor data
- tank level data
- gyro and GPS data
- tide gauge data (not implemented)

The LMP used this data to:

- set tank loads
- define the measured waterplane
- check the location of the starboard sponsons
- check the buoyancy of the starboard sponsons
- evaluate stability
- evaluate shear forces, bending moments and torsion

7.1 WATERPLANE DATA

Measuring the waterplane was similar to doing a real-time draught survey throughout the operation. In this case, the *measured* waterplane was continuously compared to the *calculated* waterplane as discussed above.

There were 30 draft sensors installed in pairs on the starboard side sponsons. There were 14 draft sensors installed in pairs on selected port side sponsons. All draft sensors were pre-defined as *critical points* in the software so that the depth of any one of the sensors was available to set the depth of the waterplane.

In addition to the draft sensors, the casualty was divided into five sections in order to visualize the relative deflection of the hull. A gyro was installed in each of the five sections to provide heel (roll) and trim (pitch) of each section. A GPS unit was also installed to provide position and elevation data of each section. Third party software used this data to quantify hull deflections about any axis.

Providing there was no significant deflection of the hull, the *measured* waterplane would be defined by the depth of a single draft sensor along with heel and trim of the casualty provided by one of the three gyros. Using depths from two more draft sensors and heel from one gyro would allow the waterplane to be defined using a straight line.

If hull deflection about the transverse axis (hog/sag) was significant, the user had the option to use multiple draft sensors to define the waterline by a parabolic curve. Manual contingencies were in place to deal with significant hull deflections about the vertical axis by bending the geometry file in third-party software.

Manual contingencies for measuring and inputting the data necessary to define the waterplane were put in place in the event of a failure of the CMS.

By default, the reference line for all drafts was the baseline of the wreck. This was a straight line along the longitudinal axis of the wreck. The LMP could be configured to use a reference line other than the baseline to account for the chains, bilge protection plates and blister tank structural members all of which extended below the baseline. All drafts, input or output, measured or calculated, would then refer to this user-defined baseline.

Tide levels affect a grounded vessel as lower tides increase ground reaction and vice versa. The LMP allowed the user to raise and lower the waterplane to simulate the change in tides. The rise and fall of the tide could be input manually or provided real-time from a tide gauge. Tide levels do not affect the stability or residual structural strength of a floating vessel and were not required once the wreck was afloat.

7.2 TANK LOADS

Each sponson tank was fitted with four pressure sensors: two at the top and two at the bottom. The LMP used data from these sensors to calculate the water level in the tank. The two sensors at the top of the tank measured the air pressure in the tank. The two sensors at the bottom measured the total pressure at the bottom of the tank. The level in the tank was calculated from the difference between these two pressures.

All sensors, both draft sensors and tank level sensors, were installed in pairs for redundancy. Sensor accuracy was checked on each *idle* by comparing the readings of any pair of sensors as follows:

- The first check was to ensure that a particular pair of sensors were working. If any sensor was not working then the LMP would receive a reading from the CMS of "999" for that sensor. If the A sensor was "broken" then the B sensor was used and vice versa.

Audible and visual alarms resulted when sensors were broken.

- The second check was to ensure that the two sensors were sending the same pressure reading within user-defined tolerances. If the absolute value of the difference between the readings of the A and B sensors was more than a preset tolerance then alarms would be raised and neither of the sensor values would be accepted until a manual check of the air pressure in the tank could be made to determine which sensor, if any, was correct. The manual check was taken from analog gauges fitted on the sponson manifold.

As long as at least one sensor at the top of the tank and one sensor at the bottom of the tank were working then the level in the tank could be determined. This level was read into the LMP which then calculated the weight and centre of gravity of the seawater in the tank at the given trim and heel. If both sensors at the top were broken then the level in the tank could not be detected automatically. The same was true if the two bottom sensors were broken. In this case, the air pressure at the top of the tank could be taken from the analog gauges on the manifold. The depth of the seachest valve could be estimated and used to calculate the pressure at the bottom of the tank. The difference between these two values could be used to calculate the level in the tank providing the seachest valve was open and the pressures in the tank were at equilibrium.

Tanks without sufficient sensors to determine the water level could be changed to *bubble* tanks in the LMP with the *reference point* for the tank set to the height of the seachest valve. The manual pressure readings could then be taken at the manifold and entered into the LMP for that tank. The LMP would automatically calculate the water level in the tank providing the seachest valve was open and the pressures in the tank were at equilibrium.

Tank loads were shown with contents represented by different colours depending on the situation in that particular tank. The contents by colour were as follows:

Ballast	Blue	User-defined
Blowing	Green	User-defined
Filling	Cyan	User-defined
KEEP EMPTY	Black	User-defined
KEEP FULL	Dark Blue	User-defined
Hanging	Yellow	User-defined
Inactive	Grey	User-defined
Bubble	Magenta	User-defined
Broken	Red	Auto

All of these colours represented sea water with a specific gravity of 1.025.

7.3 SPONSON BUOYANCY CHECK

In the first phases of the refloating operation some of the starboard side sponsons were not attached to all of their chains and could not take the maximum loads they were designed for. Sponsons S1-S15 and P1, P2, P14 and P15 required positive or negative buoyancy at all times due to the nature of their connection to the wreck.

A check was done automatically on each *idle* to ensure that each of these sponsons had sufficient buoyancy. This check was based on the tank's content name. The content colour was set for ease of reference in the graphic interface.

- If the content name was "Ballast", the sponson was fully secured by four chains and could take 100% of its design load (i.e. all tanks empty at the maximum draft). A minimum buoyancy of +100 tonnes was required to keep the sponson afloat.

- If the content name of the tank was "Inactive", not all of the chains were attached properly due to interferences along the bottom. A user-defined buoyancy range of +100 to +150 tonnes was set to keep the sponson afloat without overloading the engaged chains.

- If the content name was "Hanging", none of the chains could be attached to the sponson. These sponsons were literally hanging on hooks fitted to the casualty until the chains could be connected. A user-defined buoyancy range of -100 to -150 tonnes was required to keep these sponsons on the hangers.

Audible and visual alarms were activated if the buoyancy of any sponson fell outside of the user-defined values. This was critical as the buoyancy of a sponson could change without the user ballasting or de-ballasting tanks in that sponson (e.g. with changes in the waterplane).

8. LMP OUTPUT

The LMP provided real-time output data and calculated results to the CMS on each *idle*. The CMS provided real-time visual display of the output data and results on any one of the eight 60" monitors in the ROC.

The following data could be plotted over a user-defined time period.

- Heel, Trim & Draft Data

- Ground Reaction - This file included calculated values of the total ground reaction and the longitudinal and transverse centres of reaction.

- Correction Factor - This file included calculated values of the *CFT* and its longitudinal and transverse centres of gravity.

It was extremely useful to see how these values were trending over time to ensure that the operation was progressing in the right direction. Other real-time output data that was provided in graphs included:

- Waterline
  - Traditional 3-point plot with hog/sag
  - Straight line fit without hog/sag
  - Parabolic curve fit with hog/sag.

- Righting Arm Curve
  - Plotted values of righting arm vs. heel angle from 60° port to 60° starboard. This showed a significant range of stability for the casualty for the given condition.

- Shear Data
  - Plotted values of shear force at user-defined locations. These calculated values were plotted against permissible values.

- Bending Moment Data
  - Plotted values of bending moment at user-defined locations. These calculated values were plotted against permissible values.

Other output files were automatically created with a unique date-time stamp in the filename during each *idle*. These files were saved to the CMS server and included:

- *yyyymmddhhmmss*.DAT

This data file saved the actual pressures being reported by the CMS, the checks on the data and how the level was calculated for each tank. These files were used to track errors in the input/output files.

- *yyyymmddhhmmss*.CND

This condition file stored all of the information required to define the loaded condition of the casualty. It allowed the user to go back and restore the condition that existed at the time the file was saved. Saving these files on each *idle* provided a historic record of the operations.

- *yyyymmddhhmmss*.GF

The LMP also saves the geometry file *yyyymmddhhmmss.GF* that corresponded with the *.CND file of the same unique filename. This was necessary as the geometry file could change from one condition to the next (i.e. by moving starboard sponsons).

9. CONCLUSIONS

The Load Monitoring Program was substantially more than a typical loading computer for an intact vessel and was, in fact, much more than a salvage engineer would put together for a typical refloating of a grounded vessel with flooded compartments and structural damage. The Load Monitoring Program was well integrated into the Control & Monitoring System and provided real-time results for the evaluation of the actual operation versus the planned operation. It allowed the salvage team to minimize shear forces and bending moments throughout the operation, particularly in way of the damage areas, and to deliver the vessel at contractually mandated drafts.

10. ACKNOWLEDGEMENTS

The author would like to acknowledge the entire M/V COSTA CONCORDIA wreck removal team, in particular Inken Früling and Tim Habekost for their tireless efforts in configuring, testing and operating the LMP and David Fieuw for his expertise in helping to integrate the LMP within the CMS. The author would also like to acknowledge the technical support team at Creative Systems, Inc. for their assistance during the development of the LMP, including numerous modifications to the GHS software.

11. AUTHORS BIOGRAPHY

Phil Reed is the Managing Director of Reed Maritime LLC and is an active consultant to the salvage and wreck removal industry. Mr Reed worked for over 20 years with Titan Salvage where he was involved with well over 100 salvage and wreck removal projects as naval architect, project manager and salvage master. Mr. Reed was intimately involved with the wreck removal of the COSTA CONCORDIA from bidding for the contract right through to the redelivery in Genoa. One of his many responsibilities included oversight of the Control & Monitoring System and, in particular, the Load Monitoring Program for the refloating and redelivery of the casualty.

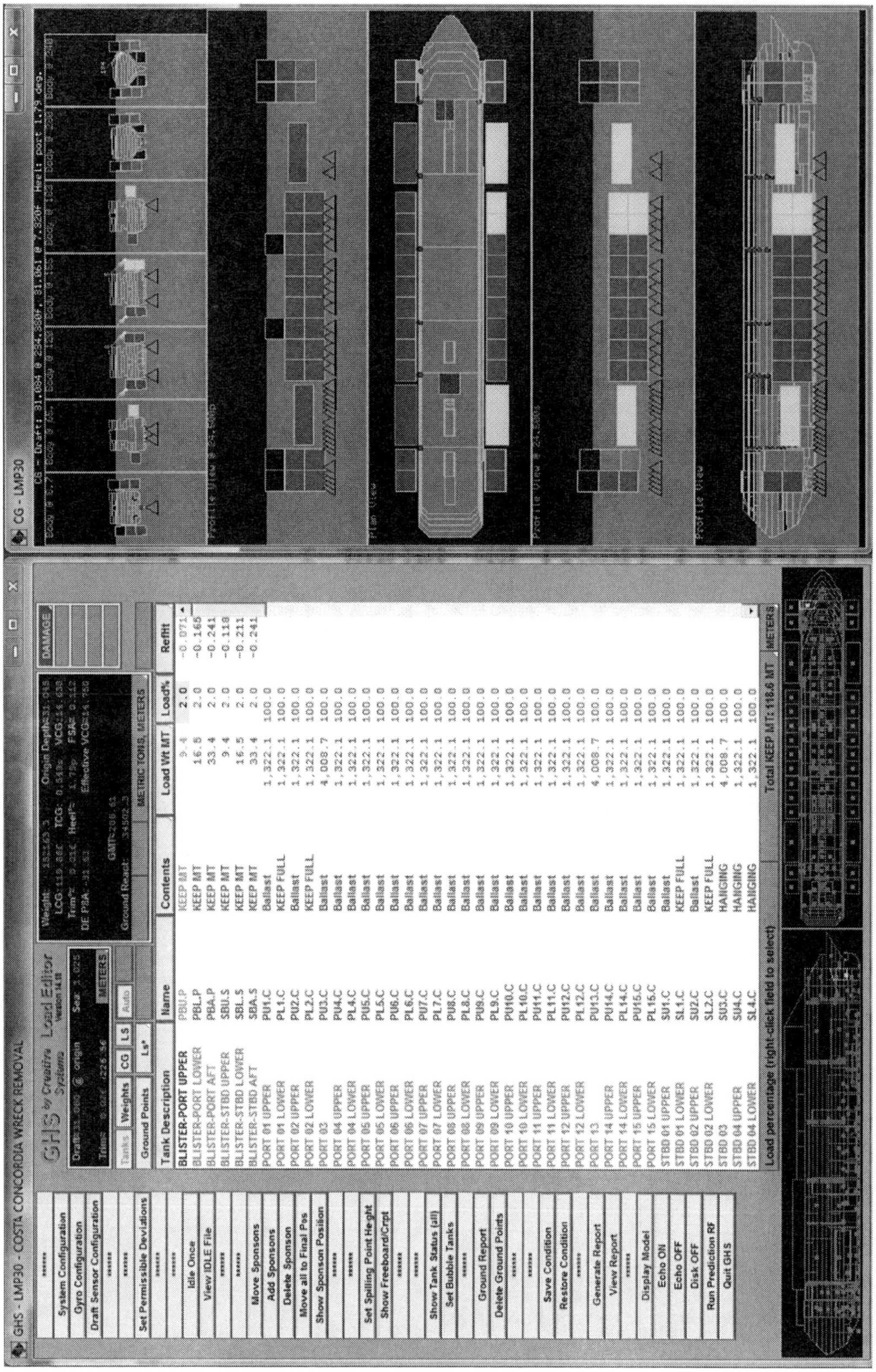

Figure 1: GHS LE Module as the graphic user interface for the LMP.

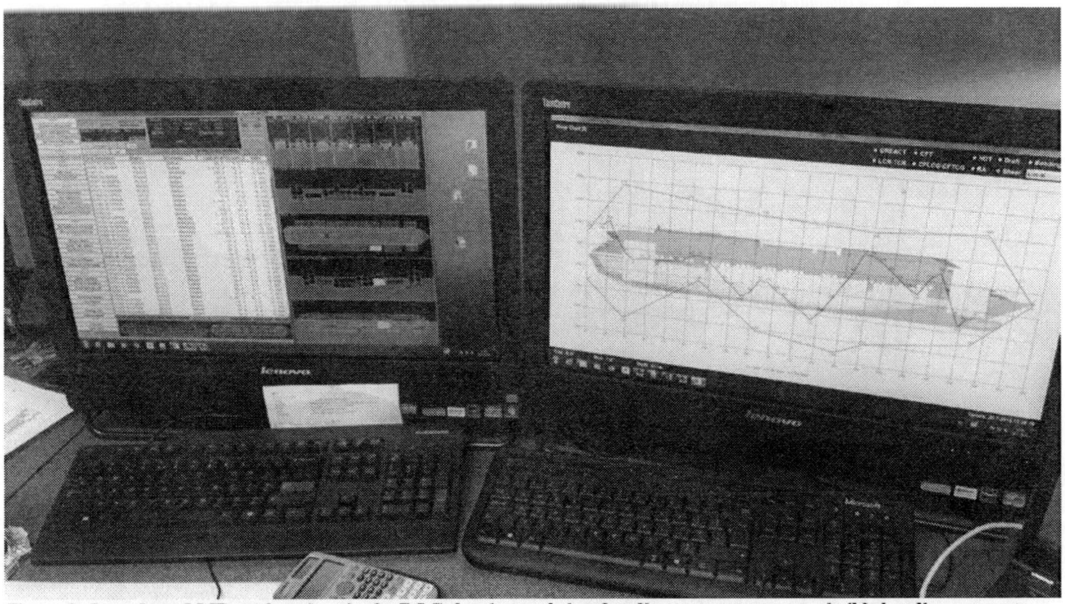

Figure 3: One of two LMP workstations in the ROC showing real-time bending moments vs. permissible bending moments.

Figure 2: Display monitors in the ROC

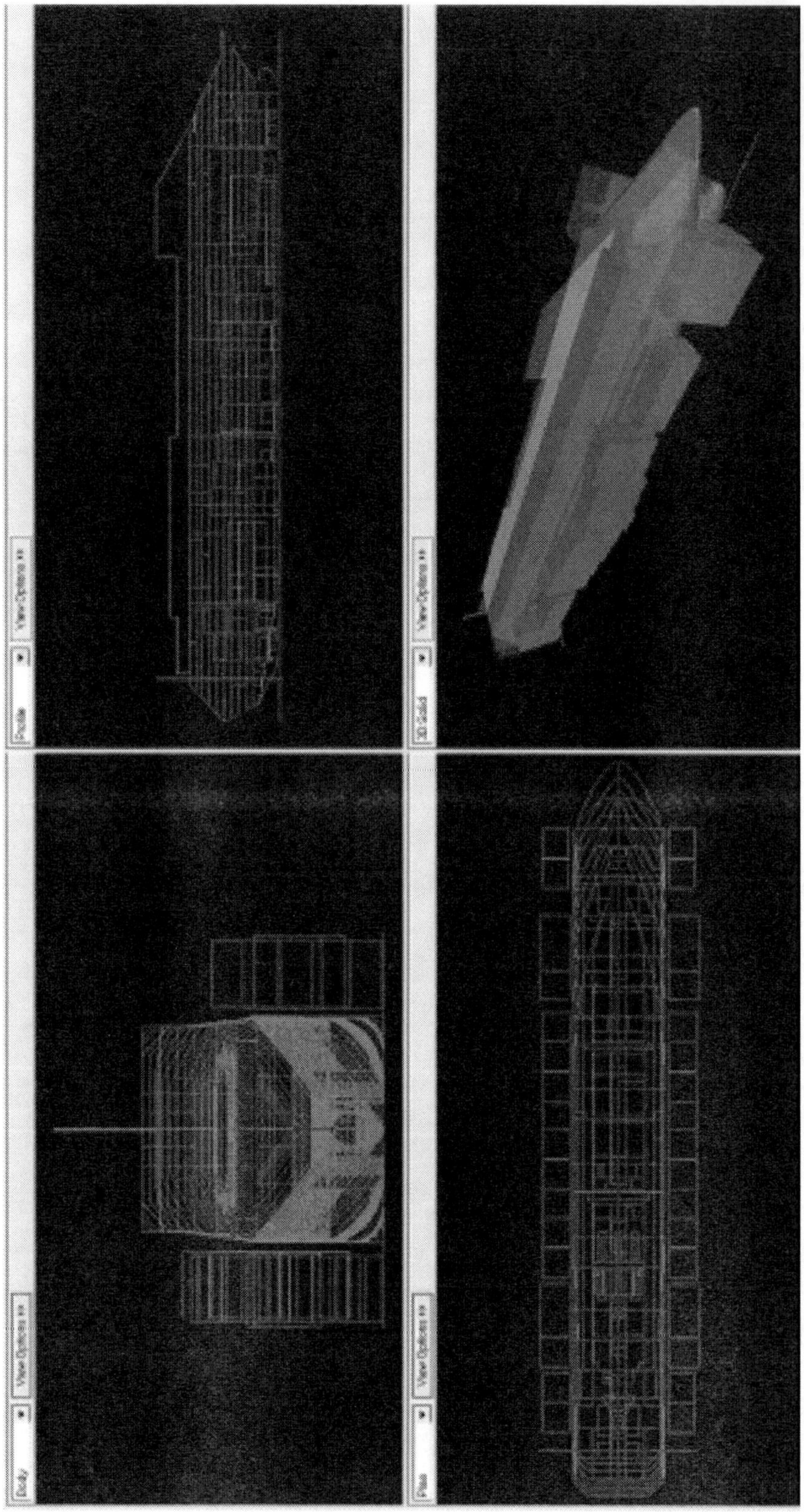

Figure 4: Geometry file - Views of the geometry showing CC hull, sponsons and 300^{+} internal compartments

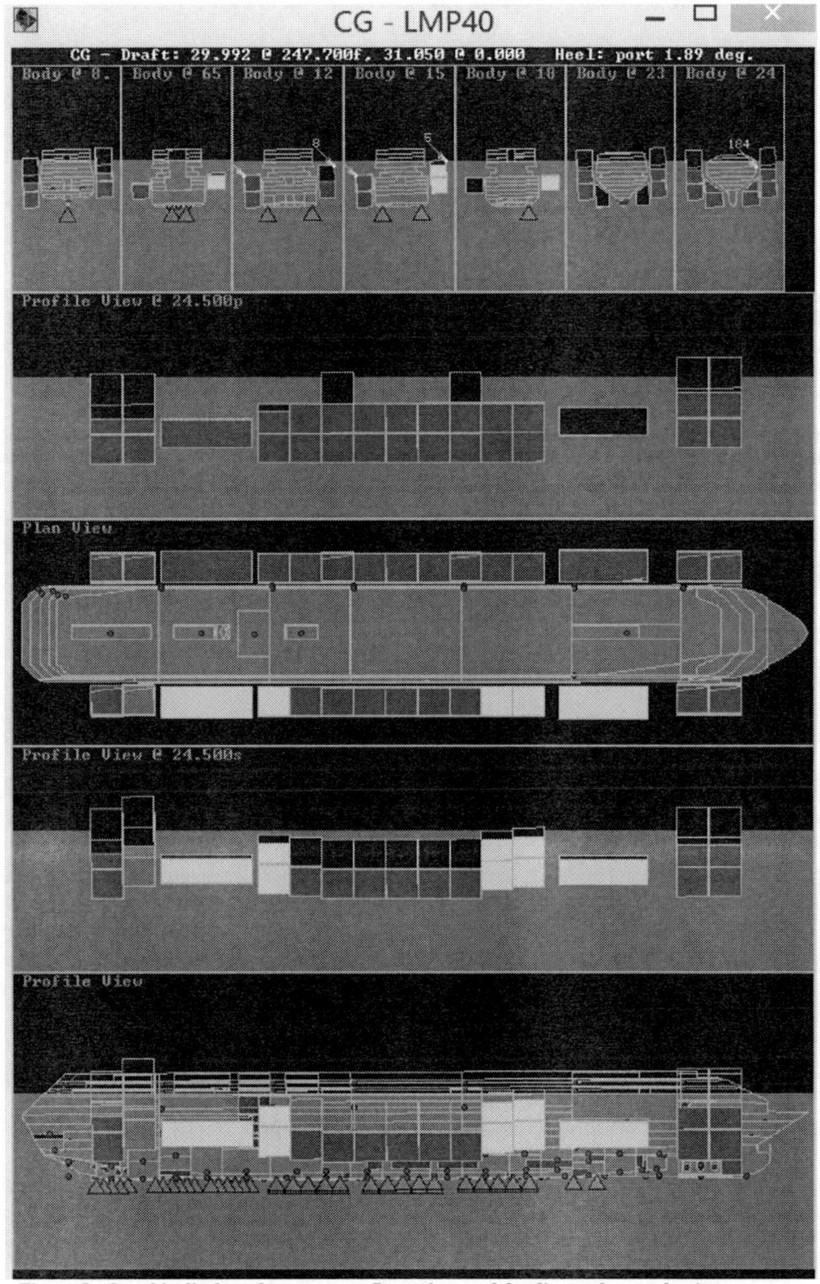

Figure 5: Graphic display of sponson configuration, tank loading and ground points.

THE DAMAGE STABILITY CALCULATIONS EVOLUTION FROM THE CAD POINT OF VIEW

Rodrigo Perez, SENER, Ingenieria y Sistemas, Spain

SUMMARY

Damage stability, as part of naval architecture calculations, has existed for many years, probably since Pierre Bouguer published in 1746 the first treatise of naval architecture, *Traité du navire.*
Historically, damage stability calculations for ships relied on manual calculations, often tied to a specific system of measurement. Some of these very old equations continue to be used in naval architecture books today. However, the advent of the towing tank and *CAD* Systems allows much more complex analysis.
Damage stability calculations are much more complicated than other naval architecture calculations, as for example, intact stability, launching or power estimation. Software utilizing numerical methods are typically employed because the areas and volumes can quickly become tedious and long to compute using other methods.
Stability tests in damage conditions, as in shipyards, technical offices or universities, are performed with software packages that starting from conceptual design information are able to quickly compute the required data. This paper focuses on evaluating old and new methods of damaged stability calculations in the shipbuilding *CAD* tools.

NOMENCLATURE

CAD	Computer Aided Design.
CoG	Centre of Gravity (m).
GM	Permissible metacentric height (m).
GZ	Righting arm (m).
HARDER	Harmonization of Rules and Design Rationale.
IMO	International Maritime Organisation.
KG	Permissible height of the CoG (m).
MSC	Maritime Safety Committee.
OSV	Offshore Service Vessel.
PSV	Platform Supply Vessels.
Ro-Ro	Roll On-Roll Off ship.
SOLAS	International Convention for the Safety of Life at Sea.
SPS	Special Purpose Ships.

1. INTRODUCTION

Damage stability, as part of naval architecture calculations, has existed for many years. Historically, damage stability calculations for ships relied on manual calculations, often tied to a specific system of measurement. Some of these very old equations continue to be used in naval architecture books today. However, the advent of the towing tank allows much more complex analysis.
Damage stability calculations are much more complicated than other naval architecture calculations, as for example, intact stability (Lewis, 1988). Software utilizing numerical methods are typically employed because the areas and volumes can quickly become tedious and long to compute using other methods.
Stability tests in damage conditions, as in shipyards, technical offices or universities, are performed with software packages that starting from conceptual design information are able to quickly compute the required data. This paper focuses on evaluating old and new methods of damaged stability calculations in the shipbuilding *CAD* tools.

2. HISTORY OF THE DAMAGE STABILITY REGULATIONS

Ship severity practical applications started around mid-18th century when the concept of metacentric height and righting arm were introduced for the first time by Pierre Bouguer, in a picture in figure 1, who published the first treatise of naval architecture, *Traité du navire*, which among other achievements first explained the use of the metacenter as a measure of ships′ stability.
It is worth it to mention that the first true stability criterion, based on metacentric height and freeboard height, was developed by Fredrik Henrik auf Chapman during the same period.
Chapman is credited as the first person to apply scientific methods to shipbuilding and is the author of *Architectura Navalis Mercatoria* in 1768.
It was not until the end of the 19th century when the first damage stability regulations was defined on the *International Maritime Conference* in Washington DC (USA) in1889.
In the article seven of this Conference it was stated:

"Ocean-going steam-vessels which carry passengers should be additionally protected by having efficient bulkheads, so spaced that, when any two compartments be filled with water, the vessel will still remain in a seaworthy condition, and two at least of the amidships bulkheads should be tested by water pressure to the height of the deck next above the water-line".

Figure 1. Pierre Bouguer (1698-1758) by Jean-Baptiste Perronneau. Musée du Louvre

During the 20th century, several International Conferences have been celebrated as regards damage stability. In the following *table 1*, a brief description of each of the conferences has been provided.

NAME	DESCRIPTION
SOLAS 1914	Signed by 13 countries on January 1914. Ratified only by five. Eventually interrupted by WWI and did not enter into force as intended on July 1915.
SOLAS 1929	New floodable length, compartment and subdivision requirements. Asymmetric subdivision to the satisfaction of the Administration.
SOLAS 1948	Minimum damage extent: longitudinal: 3.05 m ... 10.67 m; transverse: B/5; and vertical: top of double bottom to margin line. Deterministic criteria for maximum heel of 7° and no margin line immersion at final equilibrium.
SOLAS 1960	First major achievement by IMO. Signed on June 1960. Entered into force on May 1965
SOLAS 1974	A new type of convention, a convention which is amended instead of being replaced. Entered into force on May 1980. *SOLAS 1974*, as amended, is what we still use today.
SOLAS 1990	Resolution *MSC.12(56)*, adopted in October 1988. Entered into force April 1990.
SOLAS 1995	7 northwest European countries concluded a regional agreement, Stockholm Agreement
SOLAS 2009	Probabilistic damage stability requirements for both cargo and passenger ships.
SOLAS 2020	*MSC* adopted the updated regulation in Resolution *MSC.421(98)*. Restores the changed definitions on how to measure draught and trim in *SOLAS*.

Table 1. Some CAD products which manage damaged stability calculations & its industrial supplier

It is well known that most of this conferences raised after a serious accident on the Sea. Some of the accidents are listed below:

- **Titanic(1912)**. Crashed with an iceberg during her maiden voyage from Southampton to New York and sank in the North Atlantic on April 1912. 1,500 persons lost their lives.
- **Empress of Ireland (1914).** Crashed into by SS Storstadt during thick fog in the Saint Lawrence River, Canada. Sank in the early hours of May 1914. 1012 persons lost their lives. No time to close watertight doors and the water entered through open portholes, some of them close to the waterline.
- **Andrea Doria (1956).** Crashed into by SS Stockholm outside Massachusetts. Sank on July 1956. 46 persons dead. Stayed afloat over 11 hours after the incident.
- **Herald of Free Enterprise (1987).** A Ro-Ro ferry that capsized just after leaving port. Sank on a sandbank in the evening on March 1987. 193 persons lost their lives.
- **Estonia (1994).** A ferry who capsized after the bow visor broke off in strong weather. Sank on September 1994 in the Baltic Sea. 852 persons onboard lost their lives.
- **Express Samina (2000).** Another ferry that crashed against the reefs in the Aegean Sea. Main damage above the waterline, but a stabilizer fin cut through the hull below the waterline. Watertight doors open and capsized and sank on September 2000. 81 persons onboard dead.
- **Costa Concordia (2012).** A cruise ship capsized after hitting a rock off Isola del Giglio, on the western coast of Italy. The evacuation took over six hours and of the 3,229 passengers and 1,023 crew known to have been aboard, 32 died.

A special mention, in this brief summary of the damage stability regulation history, it is for Jaakko Rahola. He published his doctoral thesis, at the *Technical University of Finland*, on May 1939. Entitled *The Judging of the Stability of Ships and the Determination of the Minimum Amount of Stability*, he especially considered the vessels navigating in Finnish waters, presenting the minimum stability using curves:

- Upper curve – the GZ curve.
- Lower curve – the dynamic stability curve (integral).

3. CAD HISTORY IN THE DAMAGE STABILITY CALCULATIONS

It is possible to track the history of the damage stability calculations, following the history of the shipbuilding *CAD* Systems.

Shipbuilding *CAD* tools help to the Marine Industry in the design and construction of ships, boats and offshore platforms.

Some authors mention that the birth of *CAD* Systems was 350 BC, with the mathematician Euclid of Alexandria.

Many of the postulates and axioms used by these *CAD* Systems are based on Euclidean geometry.

If a story line continues, there is no return to innovate in these areas, up to 2,300 years after the death of Euclid, where the concept of *CAD* defined and linked first to computers.

Most authors claim that the French engineer Pierre Bezier (see in *figure 2*) is the inventor of the *CAD* Systems at the *School of Arts et Métiers ParisTech Engineers*. This engineer developed the fundamental principles of *CAD* Systems with the program *UNISURF* in 1966. Since then, *CAD* Systems have been introduced in the academic world, as universities, and in the Industry, as well as in government sectors.

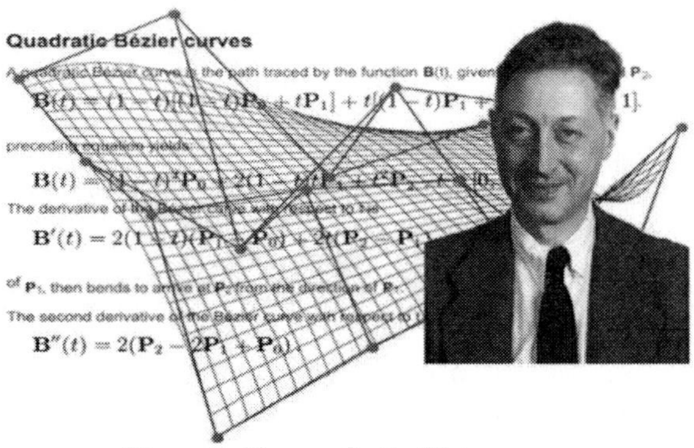

Figure 2. The quadratic Bézier curves

The real breakthrough in *CAD* Systems came under the development of computers, and it is at the beginning of 1970´s when the use of the shipbuilding *CAD* Systems becomes popular in the shipbuilding Industry. The increased use of more powerful and easy to use computers and workstations, have facilitated in the development of more intelligent shipbuilding *CAD* Systems as well as their use in day to day engineering worldwide.

The first tools to evaluate damage stability were developed around 1970´s, including flooding conditions and damage stability evaluation for defined loading conditions (see an example in *figure 3*).

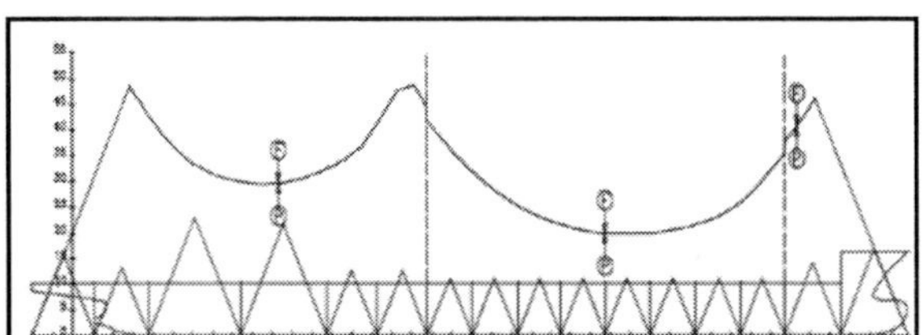

Figure 3. 1970´s output example of floodable and permissible length curves (Perez & Muñoz, 2015)

Damage stability calculation is a subject for continuous increasing need. This the reason to have continuous improvements and upgrades on the *CAD* tools, as for example this is the main cause to have the *International Maritime Organisation* (IMO) probabilistic damage calculations, probabilistic damage assessment to handle new rules for cargo and passenger vessels, calculations for warships and other functionalities/ calculations which are carried out by most of the *CAD* Systems. As a summary, in the *table 2* it is shown some *CAD* products which manage damaged stability calculations and also its industrial supplier.

PRODUCT NAME	SUPPLIER
FORAN	SENER (Spain)
Wolfson Unit Marine Design Software	Wolfson Unit, University of Southampton (UK)
Paramarine	QinetiQ (UK)
SEASAFE	Seasafe Marine Software (UK)
PIAS	Sarc (Holland)
DELFTship	Delftship Marine Software (Delft University of Technology, Holland)
ShipShape	Konsberg (Norway. Initially developed by Marintek & Trondheim University)
Hypet Series	Hypet Data AS (Norway)
Sesam HydroD	DNV-GL (Norway-Germany)
NavCad	Hydrocomp Inc. (USA)
Maxsurf	Bentley (USA)
GHS	Creative Systems Inc (USA)
Autoship	Autoship Systems Corporation
VERISTAR STABILITY	Bureau Veritas (France). Joint development with Conmar (Croatia)
Sea Solution	Sea Tech (Russia)
MAAT HYDRO	Sistre

Table 2. Some CAD products which manage damaged stability calculations & its industrial supplier

It is also well known that there are two applicable damage stability calculation methods for vessels: deterministic and probabilistic damage stability. The correct/applicable method depends on the ship to be analysed. The probabilistic method, which is based on damage statistics in terms of the magnitude and position of the damage on previously damaged ships, is considered to be more reasonable than the deterministic approach.

The improvement of probabilistic damage stability regulations started in the late 1960s, and almost a decade later, Resolution A.265 (VIII) of *SOLAS 1974* entered into force. This approach was an alternative to the deterministic procedure for dry cargo and passenger ships, which still was a fragment of the *SOLAS*.

Today's prevailing probabilistic damage stability regulations can be found in *SOLAS Chapter II-1, Part B-1*, and stems from a project named *HArmonization of Rules and DEsign Rationale* (HARDER).

The *HARDER* project began to gather huge number of collision and grounding data in 2001, with the goal of harmonizing the damage stability regulations for all type of vessels. All existing methodologies to probabilistic damage stability calculations were reconsidered and new mathematical formulations were suggested to the existing regulations. The final proposals were adopted in 2005 by the *Maritime Safety Committee* (MSC), an underlying committee of *IMO*, and the new regulations come into force in 2009.

These are relevant for dry cargo ships with a length above 80 meters and passenger ships built after 1st of January 2009. With the new probabilistic damage stability regulations in place, *IMO* actually declared that deterministic damage stability has no future (Vassalos, 2014).

Due to the *Code of Safety* for *Special Purpose Ships* (SPS), which was adopted for first time in 2008 by *IMO*

Resolution *MSC.266(84)*, SPS are also covered by the probabilistic damage stability regulations.
If a vessel has *SPS notation 1*, it is accepted as passenger ship in accordance with the damage stability regulations as regards *SOLAS Chapter II-1, Part B-1*.
Platform Supply Vessels (PSV) are, for example, not covered by the *SPS Code*, which means that *PSV* are excluded from the probabilistic damage stability regulations. *IMO Resolution MSC.235(82)* must then be applied, which sticks to a deterministic damage stability approach.
Offshore Service Vessels (OSV) are classified with *SPS* notation; more than ever, *OSV* are now involved in operations that needs more on-board distinct personnel. Wind farm operation service vessels could be an example of this type of tasks.
There are particular design advantages with the use of damage stability probabilistic approach. There are also other advantages using the traditional deterministic methodology. One of them is the straightforward calculation procedure that makes it possible to rapidly estimate the ship survivability.
The difficulty of the probabilistic damage stability regulations probably indicates how realistic and accurate the probabilistic approach is. Moreover, the probabilistic damage stability regulations most of times offer flexibility in the design of watertight arrangements which the deterministic damage stability regulations cannot compromise. With the deterministic damage stability regulations, the position of watertight bulkheads is defined by pre-defined damage extents. In the case with the probabilistic damage stability regulations, watertight bulkheads can be placed anywhere (Djupvik et al, 2015).

4. DETERMINISTIC DAMAGE STABILITY CRITERIA IN CAD SYSTEMS

4.1 STATE OF THE ART

Damage stability is obviously related to the permanency characteristics of a damaged ship. In *SOLAS*, a damage is defined as the three dimensional extent of the breach in the ship (Perez & Riola, 2011a). After injure, one or several compartments of the vessel may be filled with water, depending on the damage situation or position, the three dimensional damage magnitude and the watertight subdivision of the ship.
A space, compartment or tank, is defined by *SOLAS* as an on-board space within watertight bulkheads or boundaries, where volume is defined as a combination of *rooms*. A *room* is a part of the ship limited by bulkheads and decks, with a specific permeability (IMO, 2008). A compartment is said to be exposed to flooding when it is filled with water, and the water can move freely in and out of the compartment (Patterson & Ridley, 2014).
Traditionally, the purpose of a damage stability software is the calculation of the flooding conditions and the damage stability for the loading conditions defined previously. Normally, calculations can be performed according to the added weight method or to the lost buoyancy method.

4.2 LOST BOUYANCY METHOD

The displacement will be constant along all the calculations, except when there are flooded compartments carrying liquid cargo (this may happen when the initial situation is defined by a loading condition). In this case, the first step will be the loss of weights corresponding to this liquid cargoes. Then, the software will calculate the volumes of the compartments flooded up to the initial situation, and the corresponding modifications of the buoyancy, centre of buoyancy, waterplane characteristics etc. This new obtained situation will not be an equilibrium situation, but the program will estimate the new drafts trim and heel to obtain equilibrium. The equilibrium will be obtained when the buoyancy is equal to the displacement, and the centres of buoyancy and gravity are in the same perpendicular to the corresponding waterplane.
This new estimated situation usually will not be the final equilibrium situation (but will be nearer than the previous one) and the flooded process will be performed again. The program will calculate the volumes of the compartments flooded up to the estimated situation, and the corresponding modifications of the buoyancy, centre of buoyancy, waterplane characteristics, etc. After this, the program checks the equilibrium (with the corresponding tolerances) and if the equilibrium has not been reached a new estimation and a new iteration is performed.
The process finishes when the equilibrium is reached (the difference between the buoyancy and displacement are under defined tolerance and the same occurs with the residual arm).
In each iteration, free surface correction of the compartments carrying liquid cargo and not flooded is done (when the initial situation is defined by a loading condition).
For the calculation of the stability, a similar process is performed, but with a different equilibrium checking. In this case, for every heel angle, calculations are done with this fixed heel angle. The displacement is the intact displacement or modified by the lost liquids. To reach the equilibrium, buoyancy must be equal to this displacement and the centres of buoyancy and gravity are in the same perpendicular to the waterplane, but only in the longitudinal projection (free trimming). In the transversal projection the arm between these two forces will be the *GZ* arm for this heel.

4.3 ADDED WEIGHT METHOD

First step in this method is to calculate the volumes of the compartments flooded up to the initial situation, and then the corresponding modifications of the displacement and centre of gravity.
This new obtained situation will not be an equilibrium situation, but the program will estimate the new drafts,

trim and heel to obtain equilibrium. The equilibrium will be obtained when the buoyancy is equal to the displacement, and the centres of buoyancy and gravity are in the same perpendicular to the corresponding waterplane.
This new estimated situation usually will not be the final equilibrium situation (but will be nearer than the previous one) and the flooded process will be performed again. The program will calculate the volumes of the compartments flooded up to the estimated situation, and then the corresponding modifications of the displacement and centre of gravity.
After this the program checks the equilibrium (with the corresponding tolerances) and if the equilibrium has not been reached new estimation and a new iteration is performed. The process finishes when the equilibrium is reached.
In each iteration, free surface correction of the compartments carrying liquid cargo and not flooded is done (when the initial situation is defined by a loading condition).
For the calculation of the stability, a similar process is performed, but with a different equilibrium checking. In this case, for every heel angle, calculations are done with this fixed heel angle. The displacement will be the corresponding to the final situation after flooding. If the option of free communication with the sea has been selected, the displacement can be modified for every heel angle but the final values of the *GZ* curve will be related to the equilibrium situation displacement.

4.4 INTERMEDIATE STAGES OF FLOODING

The calculation of intermediate steps of flooding can be done in two different ways. These two ways are called: Lineal and sequential methods.

4.4 (a) Further Lineal Method

The engineer/designer must define a set of decreasing distances. These distances mean the parallel distance the compartment internal flotation is going to reach in relation to the previous stage external equilibrium flotation. In fact, the programme calculates the real equilibrium if it had entered a certain quantity of liquid up to this internal flotation and this internal flotation remains without entering more sea water at all. If the space is above the flotation, this space will be emptied up to the previous equilibrium flotation.
The free surface effect in the flooded compartments is always made by means of the accurate calculation of the change of position of the centre of gravity of the liquid in the compartments.

4.4 (b) Sequential

This method of calculation requires the designer to enter, for each compartment, how many steps are going to be studied up to final external flotation and in which step of the global number of steps begin the flooding. The programme will calculate this global number of steps. For instance, if compartment *A* is going to be flooded in three *steps* starting in *stage 2*, and compartment *B* in two *steps* beginning in *stage number 1*, it means that the calculation needs four *global steps*. When one space reaches its partial maximum number of steps, the programme, in any following step, will search the equilibrium in such a way that the internal flotation of the space was the external one. For the rest of stages the programme will calculate the volume proportionally to the number of steps of the space and the density of the liquid inside the flooded space will take into account the keys of replacement selected by the naval architect. When the initial internal level is above the external flotation the emptying out also is made sequentially.
The free surface effect in the flooded compartments is always made by means of the accurate calculation of the change of position of the centre of gravity of the liquid in the compartments.

4.5 OPENINGS, DOWNFLOODING POINTS AND REFERENCE POINTS

Ship openings are those points where the watertight integrity of the ship is broken, and which are not fitted with permanent means of watertight closing.
Downflooding will occur through these points if they become immersed in an equilibrium flotation.
Openings can have the following levels of watertightness:

- Reference points: They are points defined by the designer in order to obtain only information about their final position and the heeling angle of immersion. They are alike watertight openings and therefore they have not any influence in stability calculations.
- Downflooding points: Protected openings with weathertight closure. These points will effect progressive flooding when immersed at equilibrium but do not limit the range of stability curve.
- Quick flooding points: Permanent unprotected opening. These points will effect progressive flooding when immersed at equilibrium or will limit the range of stability.

4.6 KG MAXIMUM CALCULATION

The *CAD* Systems have normally the possibility of calculating either the maximum *KG* or the minimum *GM* complying with the criteria selected by two different tools:

4.6 (a) Limiting KG/GM curve calculation.

In this case is compulsory to define a set of initial condition giving a range of middle drafts, the number of drafts for which the maximum *KG* is going to be calculated, the trim value and, optionally, a *KG* value to start iterations. The result is one maximum *KG* per draft (or one minimum *GM* per draft).

For each of the drafts to be considered, the program will calculate the equilibrium flooding condition and the corresponding stability curve for the different values of *KG*, beginning with the initial value of *KB* and never higher to the initial value of *KM*.
The program checks the results with the different sub-criteria in which the criteria selected have been split. If the flooding situation with the *KG* complies with all the sub-criteria and in one of them (at least) in a range lower than 2% that value of *KG* is considered the maximum *KG*; if one of the sub-criteria is not complied with that value of *KG* is rejected, if the program cannot find different values of *KG* closer to the real maximum *KG* or if the number of iterations for *KG* is reached, the maximum of all *KG* that complies with all sub-criteria is selected.

4.6 (b) Limiting *KG* calculation complying with stability criteria for a flooding condition whose initial situation is an intact loading condition or by *drafts + KG*.

This option makes the program to evaluate the flooding condition for different values of initial *KG* up to find the greater *KG* value complying with all the criteria or advising that the programme is unable to find that value.
The calculation method is the same as the first tool.
For both tools all the criteria selected are checked. Also, it is possible to limit the *KG* value by the margin line or reference immersion points.

4.7 WATER ON DECK CALCULATIONS

The calculations are made in two stages:

4.7 (a) First Stage

Calculation of the height of water on deck, and the point in which residual freeboard occurs. The residual freeboard is the minimum distance between the damaged *Ro-Ro* deck and the final waterline at the location of the damage without taking into account the additional effect of sea water accumulated on the damaged *Ro-Ro* deck.
If the residual freeboard is *2.0 (m)* or more, no water is assumed to accumulate on the *Ro-Ro* deck, if the residual freeboard is *0.3 (m)* or less, then height is assumed to be *0.5 (m)*. Intermediate heights of water are obtained by linear interpolation.

4.7 (b) Second Stage

Calculation of stability and criteria checking. There are two possibilities of calculation:

- The volume of assumed accumulated sea water is kept the same at all trim and heeling angles and is calculated for the height of sea water from step 1 in the ship's upright position.
- The amount of sea water is calculated on the basis of a water surface having a fixed height above the lowest point of the deck edge of the damaged compartment calculated in step 1 or when the deck edge in way of the damaged compartment is submerged, then the calculation is based on a fixed height above the still water surface at all heel and trim angles.

In both methods the accumulated water is added as a liquid load with one common surface inside all compartments which are assumed flooded on the car deck.

4.8 REFERENCE DISPLACEMENT

When the engineer/designer calculates one flooding condition, righting arms could be different according to the method used either lost buoyancy or added weight, because they are calculated with different values of displacement. Nevertheless the heeling moment must be equal in both methods, so, to make comparable the results of righting arms, it's enough to refer the heeling moment to the same displacement. In this way the same results will be obtained and both methods are comparable. This is the same thing to do with the calculation of *GM*, as it appears, for example, in (Perez & Riola, 2011b).

5. PROBABILISTIC DAMAGE STABILITY CRITERIA

Nowadays, the probabilistic damage stability calculations are applied to most of the cargo and passenger vessels according to different regulations.
The probabilistic approach for damage stability of ships was introduced by the German Professor Kurt Wendel in the late 1950s.
Latest regulation about damage stability from the probabilistic point of view is *SOLAS 2009 chapter II-1* based on *MSC194(80)*.
80th session of the *IMO*, *MSC* the working group finalized a substantial revision of *SOLAS chapter II-1 part A, B* and *B-1* aiming at harmonized damage stability requirements for all ship types except tankers performed by means of a common probabilistic method.
The draft adopted at the *MSC 80* will apply to all new passenger vessels, *Ro-Ro* and cargo ships built on or after 1st January 2009. The application date chosen will allow ensuring consistency and compatibility with other *SOLAS* chapters and finalizing the explanatory notes in order to arrive at user friendly formulation and homogenous interpretation of the revised damage stability requirements.
But there are also former regulations for cargo and passenger vessels. For dry cargo ships, and before the 1st of February 1992, there was not any regulation about damage stability, except the reference, on the 1966 *Load Line Convention*, to the vessels with reduced *type B* freeboards, which were required to survive with residual stability after sustaining damages up to a maximum defined size that was not necessarily applied to all locations in the ship.

Since February 1st of 1992 it was added to *SOLAS*, as *Part B-1* of the *chapter II-1*, a new Regulation for subdivision and damage stability of cargo ships over *100 (m)* in length that applies to the ships of this type constructed on or after February 1st of 1992. Later, the Regulation was extended to include ships with length between 80 and 100 meters.

Damage stability of passenger ships had been considered in *SOLAS* (*part B* of the *chapter II-1*) long time before the adoption of the probabilistic methods of calculation. The eight *Assembly* of *IMO*, by *Resolution A.265 (VIII)* adopted a set of probabilistic regulations of subdivision and damage stability of passenger ships as an equivalent to and a total alternative to the requirements of *part B* of *chapter II* of the *International Convention* for the *SOLAS 1960* for passenger ships.

A deterministic damage stability criterion is all about guaranteeing the safety of a ship. As it has been shown during the paper, it has been the methodology used for a long time, and the deterministic damage stability regulations are still in practise today for several reasons. Firstly because the probabilistic damage stability regulations do not cover all types of ships. Secondly, despite the deterministic damage stability regulations' well-known traditionalism, it has a long track-record, and the engineers inclines to trust on well-proven methods.

In addition to the *SOLAS 1974* standard, which regulates ordinary passenger ships, Table 3 provides an overview of other *IMO* instruments that contain deterministic damage stability provisions (Hjort & Olufsen, 2014).

REGULATORY FRAMEWORK	APPLICATION AREA
ICCL-66	Cargo ships and tankers with reduced freeboard.
MARPOL-73/78	Tankers carrying cargo oil.
IBC	Code Ships carrying dangerous chemicals in bulk.
IGC	Code Ships carrying liquefied gases in bulk.
HSC	Code High Speed Crafts.

Table 3. IMO instruments containing deterministic damage stability provisions, besides of SOLAS 1974 (Hjort & Olufsen, 2014).

6. CONCLUSIONS

CAD tools made damage stability calculations in accordance with the latest international conventions, regulations, rules and resolutions, which are continuously updated, even some *CAD* can define their own criteria.

CAD Systems enables accurate calculations of flooding conditions and damage stability according to the deterministic and probabilistic methods, taking into account the intact stability conditions, and associates the spaces to the geometric subdivision defined.

The short term future in the software development will be the calculation of the longitudinal strength after damage to know the situation after grounding, collision, etc.

One of the features currently under development is the calculation of the effect of the crossfloodings pipes between compartments. The application will be able to calculate the effect of the flooding of some compartments including the connection among them. This will allow not only to have a more realistic knowledge of the behaviour of the vessel after the damage but also to identify actions to get equalization of the ship by flooding opposed compartments. These calculations will also include, for example, equalization time.

Future versions of these tools will incorporate the analysis of dynamic behaviour of the damaged ship in order to know the response of the ship to special sea conditions. The applications will also be able to predict the ship behaviour under corrective actions by the master, so supporting him about how to deal in damage, what kind of actions would delay the capsizing, or even would avoid it.

It is recognized that the current regulations are still not perfect, there is a lot of room for improvement, so in the future it is expected to have more advanced methods such as time domain simulation, and *CAD* tools which make the calculations easier.

The future innovations of these damage stability *CAD* tools offer a hopeful future for the shipbuilding Industry.

7. ACKNOWLEDGEMENTS

We offer our regards and blessings to all of those who supported us in any respect during the completion of the technical paper.

8. REFERENCES

1. DJUPVIK, OM; AANONDSEN, SA and ASBJØRNSLETT, BE., 'Probabilistic Damage Stability - Maximizing the Attained Index by Analyzing the Effects of Changes in the Arrangement for Offshore Vessels'. NTNU, 2015.

2. EKLUND, R., & LINDROTH, D., 'Workshop N2 - Damage Stability According to SOLAS 2009'. NAPA User Meeting, 2009. NAPA Ltd.

3. HJORT, G., & OLUFSEN, O., 'Probabilistic damage stability'. DNV GL AS, 2014.

4. IMO. Resolution MSC.281(85), 'Explanatory notes to the SOLAS Chapter II-1, Subdivision and damage stability regulations, Part B'. International Maritime Organization, 2018.

5. LEWIS, E. Principles of Naval Architecture. Second Revision ed. Vol. I. Boston: *The Society of Naval Architects and Marine Engineers*, 1988.

6. PATTERSON, CJ. and RIDLEY, JD., 'Ship stability, powering and resistance'. Vol. 13. London: Adlard Coles Nautical, 2014.

7. PÉREZ, R. and MUÑOZ, JA, 'The Next Step for Damage Stability Calculations'. DAMAGED SHIP III. pp 33-42. London, United Kingdom, 2015. ISBN 978-1-909024-38-0.

8. PEREZ, R. and RIOLA, JM., 'Case study of damage stability criteria of Merchant vessels and Warships'.

Damaged Ship International Conference. 26-27 January. London, UK, 2011a.
9. PEREZ, R. and RIOLA, JM., 'Damage Stability Criteria in Aircraft Carriers'. Journal of Marine Technology and Environment, Vol. 1, 2011b.
10. RIOLA, JM. and PEREZ, R., 'Warship damage stability criteria case study', *Journal of Maritime Research*, December, 2009.
11. VASSALOS, D., 'Damage stability and survivability-nailing passenger ship safety problems'. Ships and Offshore Structures. 9(3), 237-256, 2014.

9. AUTHORS BIOGRAPHY

Rodrigo Perez Fernandez *PhD* in *Naval Engineering* at the *Technical University of Madrid (UPM)*, in Spain. He holds the current position of *Project Manager* at *SENER*. He is professor at the *UPM*.

CHALLENGES IN HYDROSTATIC MODELLING OF A CASUALTY SINKING SEQUENCE

B Jiang, London Offshore Consultants Ltd, UK

SUMMARY

The investigation of a ship casualty sequence usually involves creating a hydrostatic model and simulating a sequence of events in an attempt to best replicate the actual sinking scenario. However in most cases, there may be little evidence on which to base such hydrostatic analyses, such as a few photographs or videos to identify the floatation, trim and heel. The main challenges lie in the accurate assumptions of the potential causes to the loss of initial stability, the subsequent internal cargo / fluid movement and down-flooding / progressive flooding scenarios at each intermediate step along the capsizing timeline. In addition to a valid model, it is essential to have an experienced investigator who is able to determine the most appropriate analytical approach and to distil the many permutations of causative factors into a few simple cases which, whilst approximations, may be considered accurate enough to determine the primary cause of an incident.

1. INTRODUCTION

1.1 MARINE CASUALTY ANALYSIS

London Offshore Consultants (LOC) provide technical advice and expertise at all stages of the marine claims and disputes process including marine accident investigation. LOC personnel have worked on marine casualties of all scales, including some of the largest and most complex incidents in recent history. The experience and specialist knowledge gathered over the years enables LOC's technical investigators to give independent expert opinions in litigation proceedings and provide expert evidence at formal enquiries.

1.2 SCOPE OF THIS PAPER

LOC's involvement in casualty analysis is the main subject of this paper, which describes the general analysis procedure and the challenges. The process starts with preparing a valid hydrostatic model in the aftermath of a casualty. The casualty's sinking sequence is then examined based on information received, extensive research and deliberate assumptions. Finally it is hoped that an independent and objective conclusion is reached with respect to causation.

2. HYDROSTATIC MODELLING

2.1 GHS SOFTWARE

GHS (General Hydrostatics) by Creative Systems is an industry standard program used extensively in marine casualty investigations. The program can analyse any ship orientation without limitation of trim and heel, from upright to fully inverted.

In damaged stability assessment, the extent of damage limited by compartmentalisation, and damage in the form of hull deformations can be readily simulated. GHS computes the weights and centres of all contained fluids at any given ship orientation. Partially flooded spaces are thereby accurately represented.

2.2 INPUT DATA

The requirements for creating a hydrostatic model suitable for a casualty investigation are normally case dependent in terms of the necessity and the level of accuracy. For instance, if there is no concern about longitudinal strength then the lightship distribution may be omitted and replaced by a point mass or a line mass.

Availability of the following plans [1] is desirable as soon as the requirements for hydrostatic modelling have been determined by the investigators.

- Lines Plan or Table of Offsets
- General Arrangement Drawing
- Tank Capacity Plan and Sounding Tables
- Class approved Trim & Stability Book
- Pre-Incident Loading Condition

For general longitudinal strength assessments, confirmation of compartmentation and assessment of residual strength, the following additional plans and information are also required.

- Lightship Weight Distribution
- Mid-ship Section Drawings
- General Construction Profiles
- Deck Plans and Shell Expansion
- Physical Condition of the Hull

3. CHALLENGES IN HYDROSTATIC MODELLING

3.1 INCOMPLETE DATA AVAILABILITY

Often during a casualty investigation not all of the drawings will be available, especially at short notice. Typically, lack of data may be due to owners not possessing the lines plans which may never have been distributed by the building yard or have not been passed on from previous owners. In addition, it is often the case that the lightweight distribution is not included in the loading manual.

Over the years of practicing casualty investigation, LOC have developed an in-house library of typical hull forms and GHS models, ranging from general cargo, tanker and heavy lift vessels to jack-ups, car carriers and yachts. Alternatively, the hull form can be approximated by taking offsets from waterlines, profiles or the shell expansion instead of the lines plans.

In the absence of a lightweight distribution for longitudinal strength assessment, an approximation can be generated either by methods such as spreading the lightship according to a coffin diagram and adjusting the trapezoidal sections to match the LCG, or by a station coefficient. Here a first estimate of the parallel mid-body mass per unit length is made and point masses added, derived from published initial design formulae. Manipulation of the data can then be carried out in a spreadsheet.

Whilst the above methods are proven to be relatively effective for creating a hydrostatic model in the case of lacking input data, the accuracy of the results is to some extent compromised in terms of trim and stability calculations and generally not reliable for longitudinal strength assessment. The judgement of adequacy of such approximation is in fact case and purpose dependent. For instance, if the objective of the assessment is to determine the effect of GM reduction due to the free surface effect induced by certain slack tanks or cargo liquefaction, a similar hull form would normally suffice provided that the tanks or cargo holds are modelled accurately. On the other hand, if the casualty is predominately caused by global strength failure then a strength model is required with each frame correctly representing the hull form so as to accurately calculate the weight and buoyancy distributions.

Availability of a complete set of input data will significantly reduce the amount of time spent creating a valid model.

3.2 MODEL VALIDATION AND PRE-INCIDENT CONDITION

Prior to performing independent analysis of the casualty, the model is normally validated against a number of loading conditions from the class-approved trim and stability book. This verifies the hull hydrostatics at a number of waterlines, as well as tank capacities and centres. The International Association of Classification Societies (IACS) have set out standards and values of convergence which can be used as validation criteria to assess whether the model is suitable for use [2]. Typically, the maximum tolerance is 2% for hull displacement or internal compartment volumes and 1% for GM.

As an example, Figure 1 demonstrates the correlation in the righting lever curve (the GZ curve) achieved between the GHS model and the sample departure loading condition from the trim and stability book, in terms of GM, range of stability and the heel angle at maximum righting lever. It can be noted that the first point of inflection occurs at approximately 10 degrees list indicating the deck edge immersion angle and this is correctly captured in the GHS model.

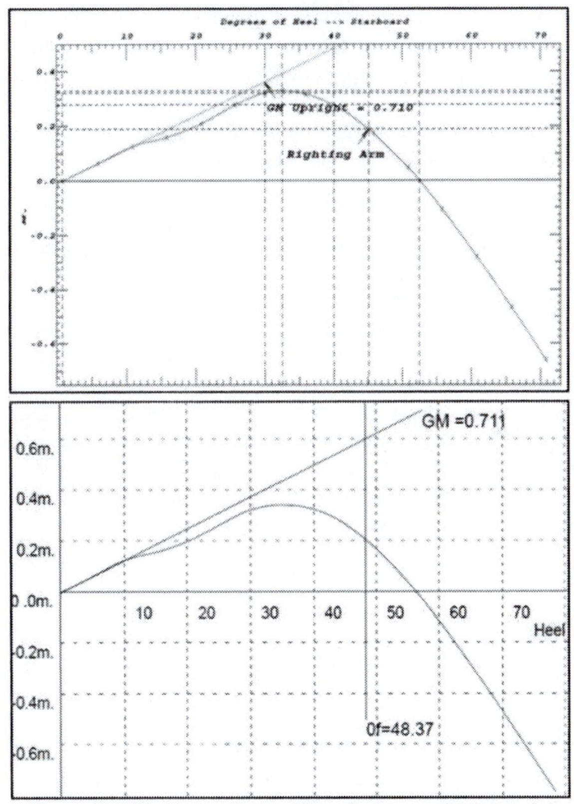

Figure 1: GZ Curve Comparison between GHS Model (Top) and Trim & Stability Book (Bottom)

When the table of offsets is available, good correlation can be expected in terms of hull form with minimum variation. For compartments however, it can be challenging to meet the IACS criteria if the ship has relatively complicated internal arrangements with oddly-shaped tanks. Modelling of such tanks will normally require input from the ship's construction drawings at typical transverse frames in conjunction with the tank capacity plan. A correct three dimensional interpretation of the drawings is essential for the model to be valid. For small compartments such as lube oil or thermal oil tanks, a general rule of thumb is to model them as point masses if their capacity is below 10 metric tonnes as the free surface moments of such tanks are minimal.

The validated model is then used to replicate the last known loading condition, usually the ship's last departure report. The input such as lightship, cargo weights and tank fillings normally require a double check against other sources such as the trim and stability book, the cargo manifest and tank sounding records etc. Speaking from experience, in many cases discrepancies are found between the load values in the departure condition calculation and other source records. Therefore, a conclusive judgement has to be made based on the various possibilities of loading patterns in order to determine the most realistic condition of the ship prior to the incident.

It is preferable to achieve correlation with the draught survey records on departure rather than the output draughts from the ship's loading computer if there is any discrepancy. If tanks are to be loaded based on ullage or sounding then the validation condition should reflect tank fill capacities rather than the corresponding mass of liquid.

When a departure condition report is not available, a presumed departure condition can be established by scaling the freeboard, the draughts and the heel angle from available photographs or video footage evidence. However, such measurements can be error-prone and in many cases not reliable due to the display aspect ratio which can distort the footage or photograph to a certain degree when compared to reality. Therefore, care should be taken when attempting to perform such assessments. Video analysis experts may be involved if the analysis is part of legal proceedings.

3.3 ASSESSMENT OF INITIAL STABILITY

The initial upright stability indicator is the ship's metacentric height (GM). GHS is able to take the free surface moments of slack tanks into account when calculating GM. The IMO Intact Stability Code requires a GM for seagoing ships to be not less than 0.15 metres at any time [3]. Non-compliance with criteria is however not necessarily related to the cause of the casualty.

Many casualties have only marginal intact GM prior to a stability incident, based on the information provided regarding departure loading condition. A low GM still suggests there is positive initial stability against capsize, but the ship is in a tender state and becomes more susceptible to external forces induced by wind and waves or internal fluid movement such as leakage of tanks. A ship in such a state may experience high roll angles which may pose a potential hazard in respect of internal cargo shifting and down-flooding through low openings in the ship's side shell or on the open decks. Progressive flooding may lead to an inevitable capsize and / or sinking.

The pre-incident loading information provided may be inconclusive in terms of actual loaded cargo or tank fillings not as described. Consequently the calculated GM even from a valid hydrostatic model may not always best reflect the actual GM of the ship at the time of the incident. Additional methods for estimating or verifying the GM have been put into practice when video footage evidence is available.

The first direct approach of estimating GM is by timing the average roll period on the video footage prior to the incident and incorporating it into the formula for roll period such as that from the DNV Ship Rules Pt3 Ch.1 B402 [4].

$$T = \frac{2k_r}{\sqrt{GM}}$$

Where,

$T = Roll\ period\ (s)$

$k_r = Roll\ radius\ of\ gyration\ (m)$

As an example, consider a fully loaded container ship with 22 metres moulded breadth berthed alongside the quay prior to departure. The CCTV footage from the berth security cameras shows the ship was rolling at an average roll period of 20 seconds. Assuming an even distribution of cargo, the GM can be estimated as 0.74 m.

The second direct approach of estimating GM is by measuring the resultant list angle due to a shift of cargo based on the video footage and incorporating it into the empirical formula for heel angle.

$$\tan\theta = \frac{w \times d}{\Delta \times GM}$$

Where,

$\theta = List\ Angle\ (deg)$

$w = Shifted\ cargo\ weight\ (tonne)$

$d = Shifted\ cargo\ distance\ (m)$

For instance, a general cargo ship was loading the last item of heavy cargo weighing 40 tonnes using the ship's own crane while berthed alongside the quay. The CCTV footage shows the ship listed to port at approximately 10 degrees from its original upright condition as soon as the crane picked up the cargo from the quay at a total out-reach of 15 metres from the centre line. Based on a ship displacement of 10,000 tonnes the GM can be estimated as 0.34 metres. However, this approximated GM takes the rise in VCG due to load on the crane hook into account. The effect must be deducted in order to derive the original GM prior to the lift.

The above approximated GM can be compared with the GM calculated from the hydrostatic model as a verification. Good correlation indicates that the loading information provided is reliable.

3.4 ACCURATE ASSUMPTIONS OF INTERNAL FLUID / CARGO MOVEMENT AND DOWN-FLOODING SCENARIOS

Once the initial GM of the casualty is determined, the investigation enters into an open-ended stage assessing various co-existing contributing factors to the loss of stability, effectively the reduction of the initial GM.

In many cases, free surface effect is the predominant contributing factor to GM reduction. This can be caused by tank leakage due to structural damage and / or piping failures, both of which are impossible to identify solely based on photographs or video footage of the ship at the time of the incident. Recommended practice is to have tanks and compartments full so far as possible in the departure condition. However, it can be a difficult task to determine whether the actual loading conditions of the tanks and compartments were precisely as per the loading plan. The only reliable evidence may be the tank sounding records which are not always accessible after the incident, especially in the case of total loss. An alternative way to verify the amount of liquids loaded is to carefully examine the sequence of events in a chronological order based on the statements or affidavit of the ship's crew following the incident. The structural conditions of the tanks and compartments are usually inspected and reported by salvors. Their reports may be helpful to indicate whether there was potential leakage in the tanks.

Free surface effect can also be developed in bunker tanks during the voyage due to daily consumptions. Therefore, in the case that the incident occurred mid-voyage, it is preferable for the investigators to have access to the daily noon reports of the ship in order to assess the effect of free surface on the ship's stability. However, in the absence of such evidence the investigators may estimate the consumption based on their previous sea-going experience and an assessment of the ship's machinery systems.

Free surface effect can take place in other forms such as cargo liquefaction of mineral concentrates or even effluent produced by loaded livestock. These case specific scenarios often require expertise in particular fields and investigators will need to engage with other experts for opinions on the matter.

Shift of cargo can result from the ship's excessive angle of heel, exacerbated by inadequate cargo securing arrangements. The assessment of the likelihood of cargo shift and the shifting distance at a certain heel angle requires detailed calculations based on the cargo type, stowage position and sea-fastening conditions. In cases where the deadweight is relatively large and the initial GM is small, a total shift of cargo even by a short distance can cause excessive list of the ship. Normally the sea-fastening equipment is inspected following the incident, as an indication of whether or not the cargo shifted during the ship's capsizing timeline.

However, in the analysis of the casualty sequence, it is often difficult to determine whether the shift of cargo is a result of excessive heel angle or the cause. Therefore, investigators will often start by assessing the effect of cargo shift on the ship's initial stability at a hypothetical level. They will consider various possibilities and compare them with other source of evidence to make a conclusive assumption. In the case that the ship develops excessive list due to shift of cargo, down-flooding may well occur through the lowest opening.

Investigators will initially identify the location of the lowest opening of the ship, based on which the down-flooding angle is then calculated in conjunction with the ship's hydrostatic condition in terms of draught and trim. The location of the lowest opening is normally provided in the trim and stability book. However, it is often that the lowest opening described as the down-flooding point is not the actual lowest point. There may be unsecured openings such as ventilators and small hatches, air pipes without correctly functioning self-closing devices and any other undocumented modifications to the ship's equipment or structural arrangements. The corresponding internal compartments connected to such openings will be assumed to take on water initially. Assumptions for further progressive flooding will then depend on the ship's watertight arrangement in terms of internal openings, watertight doors and bulkhead penetrations.

The water ingress rate through the identified opening can be approximated based on its size and submerged depth using empirical formula.

With an estimated water ingress rate, the amount of flooded water can be approximated at a certain stage along the capsizing timeline. The amount of water ingress required to induce sufficient free surface effect to finally diminish the GM to zero can also be calculated.

The main challenge in assessing a down-flooding scenario is to correctly identify the point of irreversibility. That is the stage of down-flooding at which the ship's stability vanishes. Beyond this point the situation then becomes complex due to progressive flooding through perhaps multiple other openings, eventually leading to the total loss of the ship.

3.5 COMPARISON BETWEEN HYDROSTATIC AND DYNAMIC FLOODING ANALYSIS

Hydrostatic modelling is an effective tool for creating snapshot scenarios at a particular timeframe of a casualty sinking sequence. However, it has certain limitations when compared to a dynamic flooding analysis which can account for real time flooding rates and wave response. Depending on the acceptable complexity and the purpose of the analysis, in general the hydrostatic approach has been demonstrated as sufficient for determining casualty causations in most cases.

4. CASE STUDIES

4.1 DECLARATION

For confidentiality reasons, in the case studies below I have given indicative information and results, rather than those pertaining to the actual investigation. The aim is to describe the general scope and direction of such investigations, not to identify the cause of particular cases.

4.2 CASE 1 – EFFECT OF SLACK WATER AND SHIFT OF CARGO

A multi-decked ship suffered an excessive list whilst loading cargo. A GHS model was initially prepared and validated. The pre-incident loading condition was established based on knowledge of the cargo loaded to that point and available photographs. The calculated initial GM indicated positive stability.

Two contributory factors for the list were considered to be liquids accumulating on the loaded decks and the shift of deck cargo.

Assumptions were made of the quantity and density of slack fluid on each deck based on the characteristics of its accumulation and drainage arrangements. Results showed that the effect of small amounts of slack fluid on each deck was considerable, based on our assumptions could induce a list of about 7 degrees. The GHS output is shown in Figure 2.

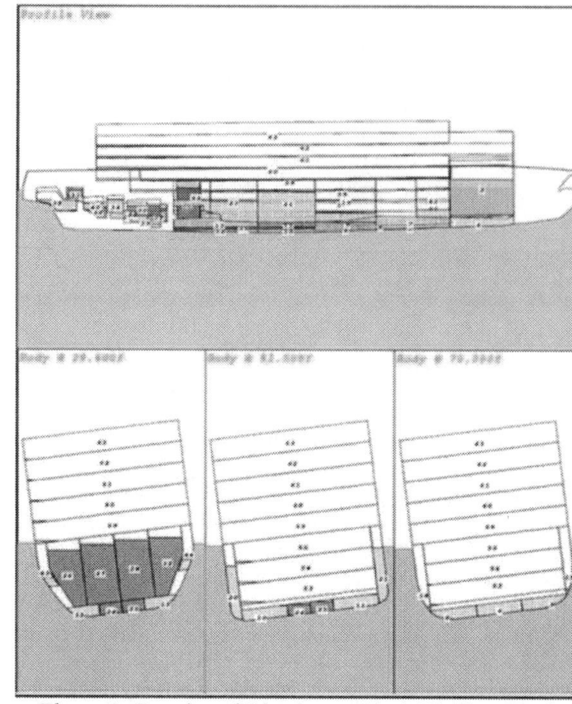

Figure 2: Developed List due to Slack Fluid on Deck

Calculations of the possible extent of cargo shift were carried out based on the weight of cargo, the propensity to move initiated by the above mentioned static heel angle and the cargo containment arrangements. The effect of a cargo shift of 0.5m was shown to further increase the list to about 16 degrees.

Meteorological records, backed up by photographic evidence, indicated that the wind strength and direction could also have contributed to the list. Assessment of the likely wind heeling moment showed it was sufficient to cause the ship to list further to about 19 degrees. The appreciable wind heel was due to the initial GM being reduced by the assumptions of slack fluid.

4.3 CASE 2 – IDENTIFYING THE CRITICAL DOWN-FLOODING POINT

A ship partially loaded with cargo that could liquefy in certain holds developed a heavy list at sea, was abandoned by her crew shortly afterwards and then eventually lost.

A GHS model was created and validated in order to investigate the potential causes of the list that developed and to determine how sufficient loss of buoyancy might have occurred in order to precipitate the subsequent loss of the ship. The initial list was found to be consistent with the effects of cargo liquefaction. Causes of the loss of buoyancy are described in the following paragraphs.

When the ship was abandoned it had a list of approximately 30° which did not increase appreciably for some time. Calculation showed that with such list, the deck edge was submerged and this was corroborated by photographic evidence.

The trim and stability book defined the critical down-flooding point to be at the main stairway on the poop deck and the corresponding down-flooding angle was greater than 60°. It is to be borne in mind that this down-flooding angle is based on the assumption that openings such as vents and hatches which would allow down-flooding at a lesser angle of list are made watertight.

At the initial list angle the main deck hatch covers were above the waterline but the sealing mechanism was considered to be vulnerable to leakage if submerged. Whilst, small openings on the side of the main deck would have been partially or completely submerged, allowing water ingress into the holds if not completely tight.

An investigation was performed using the GHS model to find out how much water would need to ingress through openings such as air pipes and small hatches on deck to immerse the main deck hatch coaming. It was found that only a few hundred tonnes of water ingress were sufficient to further reduce the GM, increase the list angle and bring the hatch coaming below the static waterline.

It is to be noted that the critical down-flooding point described in the trim and stability book usually refers to openings that have no means of closure. Water ingress can occur earlier through weathertight openings which are not designed to prevent leakage when continuously immersed at a large angle of list, although the degree of leakage may be restricted.

4.4 CASE 3 – ESTIMATING THE FLOODING SEQUENCE FROM PHOTOGRAPHS

An unmanned ship under tow suffered light contact damage on its side shell with another passing ship. Some weeks later the unmanned ship took on a list to starboard and then over a period of two days the list increased and the ship sank. The investigation was to estimate the probable flooding sequence and whether the ingress could be attributed to the previous contact.

The progress of the list and the approximate draughts during this final period were estimated from an extensive series of photographs. The list angle was measured by comparing the angle of the stem bar or derrick masts with the horizon, corrected for the point of view. The forward and aft draughts were measured from the freeboard visible at the bow and stern.

This list angle was observed to increase from about 8 or 10 degrees to more than 25 degrees over a period of about 60 hours. Shortly thereafter the list progressed very rapidly and the ship sank in a few hours.

The contact damage had occurred on the starboard side of No.2, 3 or 4 Cargo Hold and at No.3 Starboard Double Bottom Tank.

The weather at the time of the loss was not particularly severe in relation to previous weather experienced. It was assessed that the weather was not alone sufficient to cause the loss but may have contributed in causing structure already damaged to allow ingress of seawater.

A hydrostatic model was created in GHS and validated against the reported departure condition. The model profile of the casualty is presented in Figure 3.

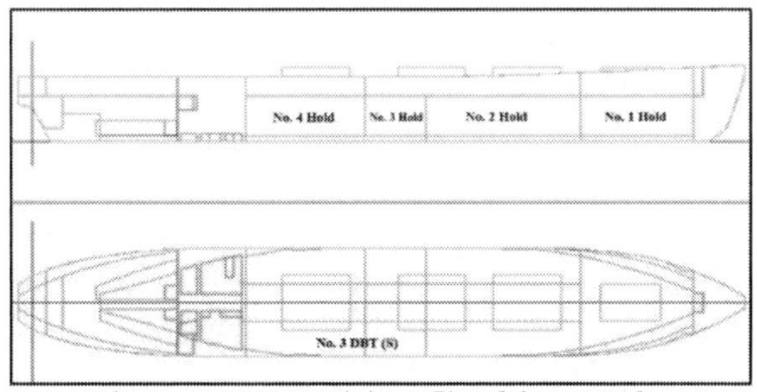

Figure 3: GHS Model Profile of the Casualty

It was ascertained that flooding of the starboard double bottom tanks alone would cause a list angle of no more than 8 degrees. Flooding of one or more cargo holds must have occurred and the investigation focused on the sequence of hold flooding that would need to occur in order to match the observed draughts and list over time.

The following sequences were considered on the basis that transverse watertight subdivision had been compromised and that flooding of adjacent compartments was likely to occur once each space was fully flooded.

- Scenario "A" – Sequential flooding starting from aft progressing forwards with Aft Peak Tank
- Scenario "B" – Sequential flooding starting in No.4 Hold then flooding forwards
- Scenario "C" – Sequential flooding starting in No.2 Hold then flooding aft
- Scenario "D" – Sequential flooding starting in No.3 DBT(S) then flooding of No.2 Hold, No.3 Hold and No.4 Hold

All four scenarios considered were thought to be possible. However, the sequence in Scenario "D" best matched the increase in draughts and list observed in the photographs, as illustrated in Figure 4.

It was recognised that a necessary approximation was that, in reality, adjacent tanks and compartments would have been filled progressively whereas in the GHS model each compartment had been assumed to flood to the external waterline before the adjacent ones.

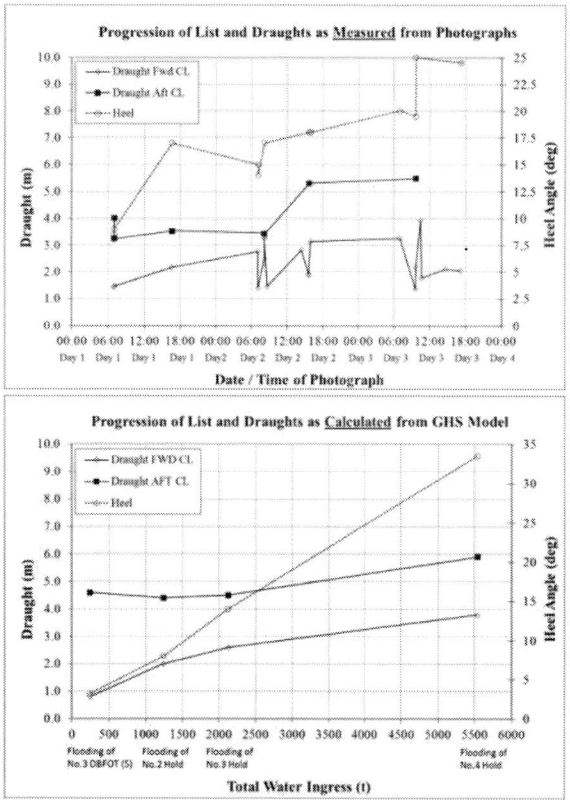

Figure 4: Comparison of List and Draughts Progression between Observed (Top) and Calculated (Bottom)

Whilst such an investigation cannot conclusively prove the actual sequence of events, it is helpful in demonstrating which scenarios most closely match the observed condition of the ship and thereby attributing a most likely sequence.

5. CONCLUSIONS

Casualty analysis with the aid of hydrostatic modelling not only requires a valid hydrostatic model but also extensive knowledge and research in many fields. The difficulties usually lie in the lack of available data for model creation and the accurate assumptions of various possible loading patterns in order to best reflect the pre-incident stability of a casualty.

More importantly, it is essential to have experienced investigators distilling the many possible permutations of potential causes to the loss of stability and the subsequent down-flooding scenarios into a few simple cases which may be approximations but are sufficient enough to substantiate or refute a particular case. Although there may be numerous amount of contributing factors to the causation of a casualty, a thorough assessment can lay out all the possibilities in the order of significance and eventually an independent and objective conclusion can be established.

6. REFERENCES

1. MANGRIOTIS, P, 'Refloating and Strength Assessment of Grounded Ship', *RINA Damaged Ship Conference*, January 2011

2. IACS, 'Computer Software for Onboard Stability Calculations', *IACS Req.Rev.3*, June 2017

3. IMO, 'Adoption of the International Code on Intact Stability', *RESOLUTION MSC.267 (85)*, December 2008

4. DNV, 'Hull Structural Design, Ships with Length 100 metres and above', *Rules for Classification of Ships*, July 2013

7. AUTHORS BIOGRAPHY

Baiqian Jiang is a Naval Architect at London Offshore Consultants Ltd where he provides naval architecture consultancy services for P&I clubs, legal firms, shipowners and hull insurers. Baiqian has been with LOC since graduating from Newcastle University in 2015 holding a Bachelor Degree in Offshore Engineering and a Master Degree in Naval Architecture. He is also an associate member of The Royal Institution of Naval Architects and is currently working towards chartership status. During the time with LOC, Baiqian has performed analyses and technical investigations as part of legal proceedings in various marine casualty incidents as well as tender evaluation for wreck removal operations. In addition, Baiqian has also been involved in marine warranty services where he provides technical reviews of project documents and operational site attendance capabilities for various large scale marine transportation projects.

PRACTICAL CHALLENGES FOR SALVORS, DEALING WITH DAMAGED SHIPS

A. Gorter, **R. Riethof**, SMIT Salvage BV, The Netherlands

SUMMARY

The following text mentions a ship, but this could be as well any floating object.

NOMENCLATURE

CoG	Center of Gravity
mton	metric ton

1. INTRODUCTION

As one of the biggest marine salvors, SMIT Salvage is dealing on an almost daily basis with damaged ships. Be it groundings, fires, collisions, structural failures and flooding.
This paper is about the practical challenges we face as salvors during an emergency - and/or wreck removal job and gives insight in the way we work and the technical issues we meet.

2. SMIT SALVAGE BV

SMIT Salvage is synonymous with total commitment in the field of marine emergency response and wreck removal. Our vast expertise is built on 170 years of experience.
We keep dedicated and highly trained salvage professionals and specialised equipment in a state of round-the-clock readiness to respond instantly to incidents anywhere in the world. Response centres are located in Rotterdam, Houston, Singapore and Cape Town.
Since 2010 SMIT Salvage is part of Royal Boskalis Westminster.

3. MARINE SALVAGE TODAY

The today's commercial salvage industry is a highly competitive market. There are 5 to 6 companies that work on a true global basis. Apart from these there are many regional companies who compete with the "big ones". And there are many local businesses/tug operators who chase for a job, when a casualty occurs in their home waters.

In general, emergency response jobs are performed on a LOF "no cure-no pay" contract (option to include/invoke SCOPIC), a BIMCO Wreckhire or any other commercial daily hire agreement.
For wreck removals, BIMCO Wreckstage or Wreckfixed contracts are often used, both lumpsum types.

An interesting aspect is that of the responsibility under the contract. Apart form the daily hire types, the responsibility lies with the salvage contractor.

4. A CASUALTY

Notices of a casualties reach us via many ways. In some cases the owner/manager takes the phone himself to call for assistance. Or his H&M and/or P&I insurance does this work for him and reaches out to several salvors.

In a lot of cases a specialist towage & salvage broker is appointed by owner's insurance. They will collect any offers for assistance and pass it on to owners.

Other routes how information arrives with a salvor are via the Lloyd's MIU Casualty Reporting Service, Coast Guard notices, GMDSS messages and of course VHF broadcasts, picked up by partners and agents.

Figure 1: Casualty on fire and listing.

Once the basic details of the casualty are known, a quick evaluation will be done to see if it is of interest to pursue.

5. GET THE SHIP'S INFO

5.1 CASUALTY INFORMATION

Now that the salvor is aware of the casualty of interest, they require as much as possible details of the ship itself and the circumstances it is facing.

With the vessel's name and IMO number, much general information can be traced via online databases like IHS Sea-web and Marinetraffic.com. With the main particulars we can check if a comparable vessels or a model of a sister ship is already available in our inhouse database to start.

Meanwhile the owners are requested to send the basic drawings like General Arrangement Plan, Capacity Plan, Stability Booklet, Midship Section and Lines Plan etc.

Also we ask for the latest Departure Condition incl. the Stowage Plan and detailed tank contents, the present drafts and list, any breached compartments etc. A last item is a request for images of the situation, as a picture tells more than a thousand words. With these basic drawings and the last known loading condition, naval architects can start to assess the situation.

The above seems a rather simple issue, but the level of preparedness among shipowners changes a lot. Some do not have any digital files at all of the ship's drawings. Meaning that personnel has to physically to run to the copy-shop, to scan the paper ones, before it can be emailed.

With a severe casualty, attention goes firstly to injured members of the crew, pollution by bunker – or cargo fuel, possible claims from charterers and/or cargo owners, attention from authorities and various other insurance interests.
The technical issues are then sometimes placed at a less important position, which might result in delays of the naval architecture review of the situation.

5.2 EMERGENCY RESPONSE SERVICES (ERS)

When owners are enrolled with a (class) Emergency Response Service (ERS) this should have been activated by now. As SMIT Salvage we have done many emergency response jobs together with the majority of the major class-ERS's and various private ERS's.

Cooperation ranges between formal keeping each other informed till sharing of detailed information between site and the ERS-office, as personnel of the last is seldom physically on location of the casualty.

Also the computer models are frequently changed out, as long as the policies of class allow this. Something that unfortunately is getting more and more difficult, due to proclaimed copyright of the design by the shipyard.
As we have the same software packages as the ERS uses, the salvor has quickly a 3D computer model up-and-running.

6. GET ONBOARD THE CASUALTY

One of the following steps is to mobilise an inspection team of salvage experts and travel to site. Once onboard they can start assessing the situation. In less serious cases, the crew might be still onboard to assist salvors and keep the vessel operational.

Onboard the salvors will introduce themselves to the crew, explaining the reason for us boarding their ship, re-assure them (post-stress) and try to make them assist us, explaining the lay-out of their vessel and typical features of it. Also activities to keep the ship operational are to be done by the crew. The purpose of re-assuring has as background that in some cases local authorities may have already done some damage, resulting that the crew is reluctant to talk or to share information about the casualty. Especially when (oil) pollution has occurred. Although salvors are not there to do any investigation into the reasons of the incident, sometimes knowing the background may help to found a solution.

In the more serious salvage cases the crew has been evacuated and the vessel is complete dead ship. Then we have to rely on the knowledge and experience of our own team to familiarise ourselves onboard and for example start the (emergency) generator to get ventilation running, get lights on in the accommodation or run the cargo cranes.

Figure 2: Onboard a listing casualty.

Inspecting a vessel in distress is not always without risks. One can imagine that the smoke of a vessel on fire is not good for your health and sometimes even toxic. Breached compartments can result in exposure of dangerous gas caused by the wetting of the cargo.

A wide variety of vessel types, each with its own specific cargo, have all their own risks and needs their own mitigating measures to manage the situation.

7. DRAUGHT SURVEYS

One of the first items the Naval Architect needs, are the draught readings. Together with the draughts on departure, a first assessment can be made of the grounding force, by calculating the difference in displacement. Here, any breached compartments come into play as well.
Readings are sometimes hard to get, when for example a layer of oil floats on the water, causing the marks to be covered on the ship's hull. Then it is better to measure the freeboard.
With the vessel partly sunk such that the waterline is above the main deck, one has to take other reference points and convert to the original locations of the marks.

Figure 4: Survey, missing draught marks.

Accuracy of a grounding force might vary a lot, if one is not able to observe the draughts accurately and/or confirm the watertight integrity of the casualty.

8. TANK SOUNDINGS

To understand the condition of the casualty, it is detrimental to know what content is where in the various tanks, cargo holds, machinery – and void spaces and other compartments.

If the crew remains onboard they can assist here to get familiar with the location of sounding points. Especially if the markings on top of the sounding tube cover have disappeared over time and one is not able to read the text anymore. Confusion starts if two tubes are close to each other, one for the double bottom ballast tank, the other for the double bottom void space.

Soundings made onboard a listing or trimmed vessel, need to be corrected according the tank sounding correction tables. With a list of more than app. 15°, the correction tables are not usable anymore. Computer simulation of the tanks can assist here to determine the correct volume inside.

On a ship that is still operational it might be possible to use the automated tank sounding system.

9. DAMAGED CARGO

A more difficult factor to be assessed is the influence of wet cargo for example caused by breached cargo holds. Weight corrections need to be assessed carefully to determine a correct loading condition, as input for the steps of the salvage plan.

9.1 BULK CARGO

For bulk cargo this seems relatively easy, if assumed that the cargo/water mixture is homogeneous.

If the hold is open to the sea, any wave and/or swell impact will wash the cargo/water mixture. And will sometimes wash the cargo out of the ship.

Every type of cargo have their own behaviour when in contact with water. Wet coal is a great example to show the volume of water occupied inside dry cargo without knowing it is there. From salvage experience dealing with assumptions about weights is a continuously process with the main goal to estimate the loading condition as accurate as possible. Original stowage factors on cargo documents are doubtful because of the impact of water and the possibility of absorption. Through a practical experiment with only a 20 liter bucket, the influence on the weight is relative easy to prove by adding water only. Fill the empty bucket complete till the top with the cargo. Add (sea)water until a sudden moment the mix will overflow. This confirmation is known as the permeability of cargo. For this specific type of (dust) coal the amount of water was 6 liter. Resulting in 30% more water inside the cargo. This effect is relevant for all type of dry bulk cargo like coal, sugar, grain etc.

Grain cargoes have the tendency to start rotting when getting wet. It might also release various gases like H_2S.

Cement in bulk has the tendency to create a crust on the outside surface when water floods the hold. We have dealt with a few cases where the outside of the cargo was hardened, but the inside still dry.

9.2 CONTAINERS

Containers contain many different kind of cargoes and as such each box will give different results when put under water or when on fire.

For small vessels you may want to make a detailed assessment, to check on the content of each box and determine the final weight. One can include the absorbed water in the cargo or reduce the weight if guttered by fire. With the latest container ships carrying thousands of boxes this is almost an undoable task.

During a casualty some years ago, a combined assessment was made, together with marine consultants, ship managers, Class-ERS and salvors, to determine the remaining cargo in the holds, the volume of Fi-Fi water in the holds and determination of the Centre of Gravity of these. This is more a calculated guess than a scientific approach. But at least there was a consensus between all parties, which prevented a lot of discussions later on.

On another case, where containers had to be discharged from the flooded cargo holds, weights of 50 to 60 mton for a wet 40ft container were not uncommon. Maximum payload is close to 30 mton for these 40ft boxes, but the record on this job was almost 100 mton!

Figure 5: Increased weight due to water absoption

This load can have significant influence of the container structure itself. Structural failure is not uncommon for these cases. Additional slings around the "belly" can avoid the heavy box breaking in two while lifting off board.

Cargo holds with cell guides increase the loading and discharging efficiency, however once a container ship is listing or severely trimmed, the axial forces on the cell guides results in problems during discharging. A solution is to cut away the cell guides to "free" the containers.

Figure 6: What is left? Remains from containers and cargo after a fire onboard

9.3 GENERAL CARGO

For any general cargo, damaged by the incident, an assessment has to be made to update the loading condition. Examples are plenty around; Logs, timber packages, cars and trucks, wind blades etc, etc.

The reduction of weight and the change of CoG of a superstructure, gutted by fire, has to be calculated and the Light Ship Weight has to be corrected for this.

Tallying of lightered cargo is always a challenge. Counting the grabs gives an idea, but how full are the grabs? Cranes are half of the time not fitted with monitoring equipment to read the hanging load.

Figure 7: What is left? Remains from trucks, trailers and cargo after a fire onboard

9.4 LPG/LNG

An aspect of these cargoes is the requirement to cool the cargo. In case the cooling water system may fail, an alternative system needs to be installed for example with submersible pumps.

10. PREPARING A 3-D COMPUTER MODEL

The model is to be checked with the data as found in the Stability Booklet and/or Capacity Plan. Focus will be on the hydrostatic particulars of the hull and geometry of the various holds, tanks and spaces.
An interesting space to model is the Engine Room. This space is normally not mentioned in the Stability Booklet, nor in any other documentation. So there is limited info on the volume and the centre of gravity. To make sure we have accurate modelled the space, we make sure we have the hull correctly modelled plus the tanks that are inside the engine room, which are actually subtracted from the initial space.

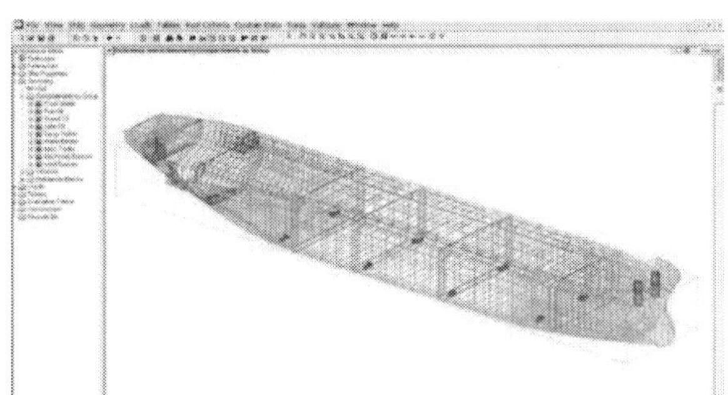
Figure 8: 3-D model of a bulk carrier in Hecsalv

For standard cases (grounding) it is sufficient to model only the hull. In case the ship is deeper submerged or capsized, it may be required to model as well the superstructure(s) and/or any cargo tanks on deck etc.

The maximum allowable strength tables for shear forces and bending moments describes the operational strength limits of a vessel. To assess the longitudinal strength the light ship distribution has to be known and added to the computer model. Data from the loading manual or docking plan can be used to define this.
The global ship structure can be analysed with the influence of damage to the structure with a simulation of transverse section(s) in a dedicated salvage software program called Section Modulus Editor or a proper XLS sheet. The position of the neutral axis and moment of inertia is adjusted for the damaged construction.

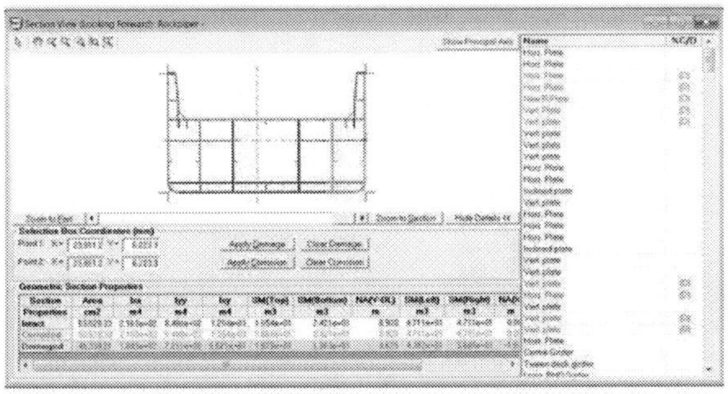

Figure 9: Transverse section of a (damaged) Rock Dumping vessel

11. THE SALVAGE PLAN

With the aforementioned information we are able to prepare a salvage plan.

11.1 Initial Condition

This is a condition for reference only, to check that the computer model is correct and to show that our loading plan is the same as used by the ship at its departure.

11.2 Present or Actual Condition

The initial condition is to be corrected for fuel and fresh water consumption and any (de-)ballasting done since departure.

Next step is the application of any damage that occurred to the hull and tanks. With the modern software it is a just a click and the compartment is damaged and it will fill or release its content till sea level. If stranded, one has to apply grounding points. There are two general ways to do this;

- Apply ground, based on water depths around the casualty,
- Apply ground, based on draft mark readings.

The second method is more often used as there is not always the opportunity to do accurate depth soundings. And it is impossible to measure underneath the hull.

One should keep in mind that the way the ground is defined is only a model of how the vessel sits aground. If for example the hull touches a rock underneath, it is impossible to say how much load or force is involved. For divers this is not possible to check, it might not even touch the hull at all, although it looks like it does.

11.3 Condition Steps for the solution

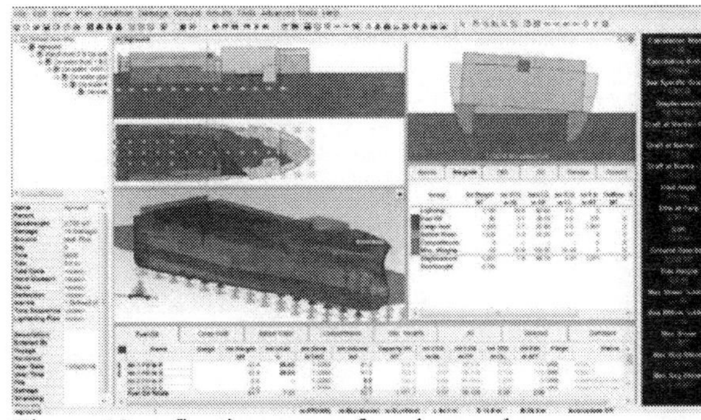

Figure 10: refloating steps of a salvage plan

11.4 Redelivery Condition

In many cases it is required, being it the owner, class, port authorities etc, to show that the vessel is a stable condition to be left without any salvage assistance. Be it for a port entry or a (long) distance towage or just at a berth.
Depending on the circumstances a suitable draft, trim and sufficient stability and strength is to arranged.

12. CONCLUSIONS

Although marine salvage is sometimes described as:

a science of vague assumptions,
based on debatable figures,
obtained from inconclusive instruments,
performed with equipment of problematical accuracy,
by persons of doubtful reliability
and of questionable mentality,

this paper hopefully has proven otherwise. But also shows that it is sometimes exactly what we are facing.

13. REFERENCES

0. HANCOX, D, 'Reed's Commercial Salvage Practice', Thomas Reed Publications Ltd., 1987.

14. AUTHORS BIOGRAPHY

Alexander Gorter is Senior Naval Architect with SMIT Salvage BV, since 2003. He is responsible for the engineering of salvage operations and wreck removals all over the world. Most salvage jobs are attended on site, to be able to act and assist the operation from first hand. Over the last years he has been acting as well as Salvage Master on various refloating and wreck removal jobs.

Roy Riethof is Naval Architect with SMIT Salvage BV, since 2013. He is responsible for the engineering of salvage operations and wreck removals all over the world. Most salvage jobs were attended on site, to be able to act and assist the operation from first hand.

DOCKING SUPPORT SYSTEM

Ahmed Samir Ghowel, MRINA, I Eng., Kuwait Oil Company-KOC, Kuwait

SUMMARY

Accidents happen, thereafter salvage takes place. On the other hand, the ship owner's decision is either going for scrapping or repairing based on the coming feedback. On the other side, the shipyard is still as it is, in a fixed site, waiting for the damaged ship to arrive.
Wasted time can be partially saved by sending drawings to the shipyard for pre-fabrication. But will be subject to martial availability and the shipyard production planning.
But by thinking even without the box, the author with his docking background, believes that the shipyard can be reconstructed by self-propelled docking and integrated facilitates that will be named hereafter F-Yard [Floating Shipyard], to serve in such an emergency case.
Based on the received data, different production process and material requesting would start. And by the time F-Yard arrives to the incident location, it will be ready for docking and repairing the damaged ship.
For vessels up to 5000 DWT, the F-Yard main function is to be utilized in case of emergency. However, it will be capable of covering class survey, in order to cover the market needs for that.
The arising problems such as lay-offs, restructuring and environment legislation ...etc. require, a balanced solution which enables adaptation of: talent management and competitive production tools, during the huge undergoing changes in its backbone structure.
Proposed approach and solutions will be illustrated as a whole docking support system. In addition, a case study of a real accident for a tug boat with a tanker in AL-Ahmadi Port.

KEY WORDS: Shipyard; Accident; Dry-docking: Heavy Lift; Floating shipyard, Agile Project Management, Lean Manufacture.

1. INTRODUCTION

From the last decades until the time being, shipyards have been profound in fixed sites that vessels had to sail to wherever they are located to perform the class survey or emergency repair. As if it is a rule of thumb.

However, the floating docking is a class approved unit, which can be tailored to sail to the required location for preforming Dry-Docking [DD]. These huge undergoing changes in their backbone structure support the working area and reduce the off hire and the risk related.

This idea was raised before in different studies and has been applied in different forms to suit the need of its target.

The difference in this paper is that the author is looking for combining different studies and designing an application that would fit for oil and gas sector at present and for serving sustainable energy in future.

This will be able to serve a wide range of vessels serving in the field. In addition to a management approach that will be tailor made for this idea.

The author's company is one of the major oil and gas companies in the Arabian Gulf. It owns a fleet of approx. 70 units [OSRV, tugs, Crane workboat, SPM, mooring, crew and oil spill barges.... etc.]. Aside to the company's fleet, there are two main contractors working locally with a total of five boats. Regardless to the other contractors that working all over the country and have their vessels as well.

For doing Class survey for this number, it takes an average of 12 days for a vessel every two years and half.

In addition to an emergency DD case. History record shows of an average of two accidents per year only in Kuwait.

The Company's fleet is serving in different oil fields all around Kuwait. These fields are producing a total three million barriers per day. This will be maximized to four million barriers per day, as per 2020 strategy. This will limit the margin of downtime.

This shows the markets need for a self-propelled floating shipyard, thereafter called F-Yard. Serving this quantity of vessels and cutting the downtime to the lowest rate.

Plus, a case study will illustrate an accident and her salvage preparation assuming that F-Yard is serving the area.

On board F-yard there are two main stations Wheelhouse and Engine Room. Wheelhouse represents the Operation Management which determines the management methodology to suit the next DD needs and constrains. While, the Engine Room represents the layout and components as a floor platform for executing the work.

Open a channel of communication, between both stations, for interaction to lean the DD operation. And for each station there are divisions of teams focusing in different particulars.

2. MARKET GLANCE

Market research and analysis needs another separate detailed study. However, there are two positive indications proving the feasibility of F-Yard.

First, locally for a country such as Kuwait, having one of the major oil companies like Author's company. Shows an average of DD 12 days per dock for each two years and half. This company is allocated in the Arabian Gulf that characterized by low taxation area and a low manpower rate compared to other areas.

For Kuwait only, number of operations expected to rise from 1492 in 2008 to 2384 in 2020, as shown in Figure [1]

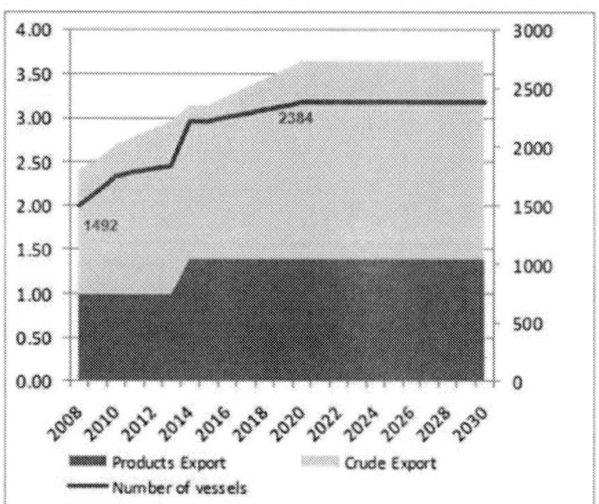

Figure [1]

Second, a global wise, it proves a business opportunity of having F-Yard serving specific area such as Arabian Gulf, Mediterranean Sea, African coasts.... etc. As its core value, is to minimize the downtime for the surrounding work boats and to maximize their revenue.

And as F-yard, acts as a multifunction unit, it is capable of working as heavy lift or a floating workshop for any industrial use, during her ideal time. Also, can be equipped by gas handling facility to match for the future of LNG harbor tugs.

For Arabian Gulf only, Lloyds Intelligence record at 2016 showed the numbers of the traffic were approx. 1974 in the Gulf calling 293 ports with a breakdown as Figure [2], X-Port, and Y- Traffic.

This reflects a big portion of the countries in the area, as part of Organization of the Petroleum Exporting Countries (OPEC).

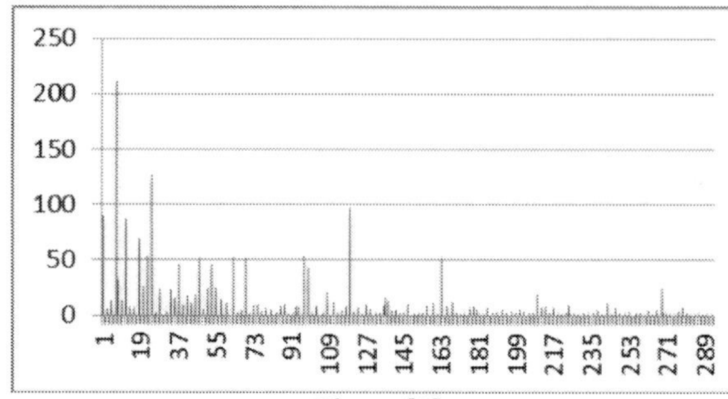

Figure [2]

3. OPERATION MANAGEMENT

The way to manage a floating shipyard in the middle of the water, for sure needs a unique way. Assuring that each step takes the right time towards the right place.

For sure the main challenge is the nature of the DD itself as an integrated chaos system.

That's the reason F-Yard adopts a management methodology that was built on top of two main focal pillars:

- Lean Manufacture
- Agile Project Management

Having Agility Lean Manufacture and having a smart Production Planning Unit.

3.1 LEAN MANUFACTURE

This target best lead time for each production process. Avoiding any non-conformity or porosity in the flow. Optimizing the process comes by eliminating production Wastes [Muda] as defined by Toyota Production System-TPS.

Having limited space became more circumspect for eliminating any wastes, from the customer's point of view, for example:

Having more material [Overproduction] than it is needed, will take up space [Inventory] which is not available. Also, like having extra [Processing] causing the material to flow [Motion or Transportation] in non-value operation and kept in hold [Waiting] exposing it to more [Defects].

Scharm [1993] in his NSRP report titled by *Non-value-added tasks for Shipbuilding* proposed a tool named "Activity Analysis" in verification of both function and cross function status.

Such tool segregates the process into three main categories, from customer perspective:

A - Value Added task
B - Non-Value-added unnecessary task
C - Non-Value added necessary task.

Then, he makes it clear that, the elimination of any task must by substitution option for continue process improvement.

Plus, Ghowel [2015] explores the other side of wastes that is generating from the customer side affecting the shipyard during:

- Project Initiation- Planning- Decision Making- Closing

Eliminating these wastes, adding an enormous saving in the shipyard physical space.

In addition to other lean tools, such as, 5S housekeeping that's essential tool when it comes for combining workshops tools and equipment.

3.2 AGILE PROJECT MANAGEMENT- APM

The reason behind selecting APM methodology, for each DD project was explained by Daniel [2009] "Agility is the ability to act proactively in a dynamic, unpredictable and continuously changing environment.

And, illustrated more by Owen et al, [2006] that, "an organizational agility is the ability to be inherently adaptable to changing conditions without having to change."

In other worlds, responding to the dynamic, changeable and uncertainty environment of DD projects, there should be a controllable, stable and fast approach like agility.

Chin (2004) gives an equation for Agile PM Environment = [Uncertainty + Unique Expertise] x Speed.

Back to Daniel, he listed down the advantages of the agility approach, which fulfilled F-Yard needs, as follows:

1- Team members collaborate in all aspects: Shipyard staff should be a high competence, multifunction expert team to be allocated in different areas.
2- Regulation of changes depends on flexible and adaptable procedures: as there is no certain scope.
3- Customer is continuously involved throughout the project lifecycle: as he is living onboard F-Yard, he is part of day to day activity.
4- Testing is well defined and done more frequently: testing at each stage is mandatory for the readiness final trials without having cost of quality.
5- Technology: by implementing pro-active innovation strategies.

3.3 AGILITY LEAN MANFACTURE

Merging the Agility and Lean in forming the operation methodology, is mainly to have an accurate, valuable and fast response. Facing unclear, changeable and interlinked scope of any emergency or normal DD.

In addition to, a cost reduction for the shipyard expenses and DD cost. Moura (2012) described it as "high level of rapid reconfiguration and will eliminate as much waste as possible".

Finally, this attracts any customer who is looking for a higher productivity and quality index.

3.4 PRODUCTION PLANNING UNIT-PPU

The Key factor for applying agility lean manufacture is the planning of its production process. At this stage, the flexibility of the F-Yard is more notable and credible than others.

The Workshops in F-Yard are based on ISO Containers, which give F-Yard the ability for having different arrangement/layout based on the coming DD requirements and optimizing the limited space of F-yard.

Lamp [2000] in his book "Engineering for Production" explores the Design for Production (DFP) approach, which is used for correlation of production design with the available or planned facilities and production methods.

This means, in case of an emergency or a normal DD, it starts by the PPU which design the full production process. Utilizing different simulation scenarios, to achieve best layout for optimizing the production flow. In respect to a risk assessment study and TPS wastes.

This could be done by using different tools to increase the utilization of the working area and eliminating the production wasted, such as: Systematic layout planning (SLP) and Analytic Hierarchy Process (AHP). As presented by Matulja [2015].

Systematic Layout Planning- SLP: A tool for arranging workplace for the areas that are high frequency and close to each other.

Analytic Hierarchy Process-AHP: A tool to organize and analyze complex decisions.

Therefore, literally, for each DD project there is a separate workshop layout and a production plan.

This flips the space restriction of F-Yard into a marketable advantage. Translated to the customer by achieving better Task Time.

4. LAYOUT AND COMPONENTS

There are different studies and applications preformed as a floating shipyard. Each one of them has its own approach.

For example, DORMAC Shipyard presents complete mobile operational platform, referred to as "Shipyard-in-a-Box", with a full spectrum of site facilities.

Other, by Boskalis through Dockwise Vanguard, that's able to carry cargoes up to 110,000 tones. She is used for dry-docking of FPSO while operation.

Another one is Suez floating dock with built-in workshops and overhead crane to cover the vessels that are passing in Suez Canal.

Other is a heavy lift ship to serve as marine Maintenance and Repair [MAR] platforms. This study was prepared by Robert Cullen [2007]. As Figure [3]

Figure [3]

This specific study is considered the core references and guideline for F-Yard.

First, F-Yard Scope of work, for vessels up to 5000 DWT at Max Beaufort force 2 [0.1 to 0.5 meters (3.9 in to 1 ft. 7.7 in)] are:

1- Provision an accommodation for shipyard staff along with crew of the repaired ship.
2- Provision of providing general service for the repaired ship such as: Water-Air-Elect…etc.
3- Provision of material storage that is required for DD and maintenance.
4- Provisions for completing Class survey such special survey.
5- Provision for preforming underwater repair.
6- Provision for conduct salvages operations.
7- Provision to transport as heavy lift vessel.
8- Provision to preform periodical maintenance of any vessel along her side.
9- Provision of gas handling system for LNG fuelled tugs, in case of LNG or Duel fuel tug.

Based on this scope, secondly, the preliminary component contents are:

A. A Heavy Lift vessel for Lifting vessel up to 5000 DWT, with open deck from both sides.
B. Rail on both sides, for 50-ton SWL crane shifting and automated Robert preservation equipment.
C. Floodable Light Modular Causeway Sections [LMCS] that is capable to lock the containers in any position.
D. Warping Tug for deploying and undeploying LMCS
E. Shelter space in form of 20 and 40 ft ISO containers.
F. Multifunction forklift with a man basket and container handler.
G. A helicopter under F-yard command.
H. A helipad on F-Yard deck and LMCS work space.

LMCS was developed by US Army R&D center; it can be stored inside a container on F-Yard deck until it is deployed on site to act as a main work space for production operation. And it will be moored/linked to the F-Yard Deck.

Containers are custom made, acting as a workshop or an accommodation. For example, machine shop, which will be equipped with multifunction machines serve different trades at single time. Number of containers and their content is determined based DD standards and normal practice. Plus, spare items for an emergency DD. All this has been calculated in a detailed study of F-yard.

DD requirements will drive the layout by deploying and mobilization LMCS and the containers.

The main use for the helipad in case of emergency injury for any personal. And to optimize the supply chain for the critical and unforeseen items.

5. CONSTRAINS

There are different constrains forming the challenging part for establishing F-Yard. However, the main ones are the environmental impact and supply chain. Others will be listed in the detailed study.

The success factor of the whole idea depends on the environment impact. This is considered the main constrain for implementing F-yard. Not only for the legislation part, but for caring about marine environment.

Aiming for a Green Shipyard to resolve this issue. Janson [2016] defined Green Shipyard as "A shipyard is considered green when the development, repair or conversion of a ship, using different processes and systems, has an environmental impact for both energy use and pollution of null."

This means environmental impact is sourced from energy use and pollution. Figure [4] draws the input and output for the shipyard operation Lifecycle that controls all process and used material.

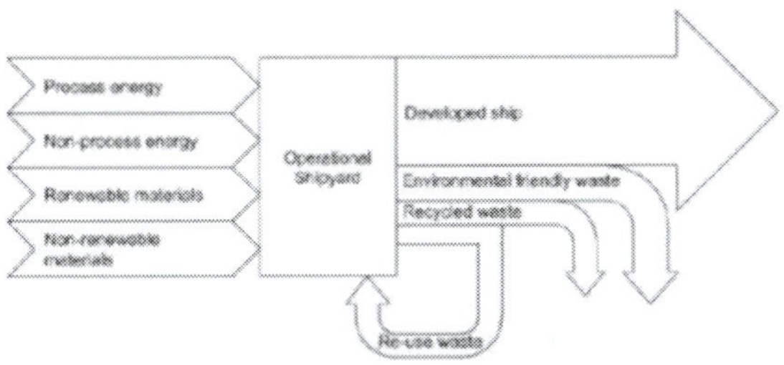

Figure [4]

Furthermore, Janson proposed different Environmental Impact Assessment schemes- EIA for "for assessment of the environmental consequence (both positive and negative) for a plan, policy, program or project."

It is vital that for each coming DD in addition to the Hazard and HAZOP for the personal and equipment. The EIA should be conducted to measure the influence of the repair on environment.

Another parameter raised by Cabral [2011], is "making the industry responsible for post-consumer disposal of the product".

This means; the F-Yard should share the responsibility for the material used during the repair. For example, using a synthetic lubrication rather than a normal one. Using a friendly grit rather than a normal one.

And looking for the future, F-Yard shall be arranged as gas handling system to encourage LNG transformation.

Another constrain is the supply chain of the material. According to Baramichail [2007], he introduced an Agile Paradigm. He stated that "An Agile supply chain is an integration of the business partners to enable new competencies in order to respond to the rapidly changing, continually fragmenting market."

In addition to, Kanban principal is the best approach for the main and consumable in general form.

Storage material is to be calculated based on the DD standard and maintenance booklets for the serving fleet. In parallel, it will be updated regularly with the coming feedbacks from the incoming reports such as measurement, readings, clearances thickness gauging …etc.

This means that F-Yard will have to build a network of suppliers for getting the material. These suppliers are getting a pull signal one material request is issued. Then after the Warping Tug will collect this material by a container, in the nearest port.

And what makes this network more valuable, is the integration business partner with F-Yard as one cooperated unit. And what makes it a Green chain, by applying the waste management by the disposal of the material for recycling after usage.

6. CASE STUDY: HAZAM ACCIDENT

On 9th September 2017 at about 0135, tug HAZAM collided with a tanker ship "DENSA ALLIGATOR" about 4 NM west of Ahmadi Port, Kuwait.

HAZAM is one of the tugs that were assigned to assist in towing the ship outside the Port.

Both vessel characteristics as in Table [1]

Characteristics	Tug HAZAM	Ship DENSA ALLIGATOR
LOA-m	38.14	224
Breadth-m	11	42
Gross Tonnage-GT	499	57219

Table [1]

The collision was resulted in a breach of 6-meter-long and 4-meter depth in HAZAM's hull port side at bulwark and extended up to one-meter under water area. Causing several indentations in the vicinity of the breach. Figure [5] shows the upper part of the damage.

Figure [5]

The collision categorized as a serious accident and required an emergency docking.

Assuming that F-Yard is serving in the area and received Mayday Call. The response action starts by the inspection team flying towards the scene, for, first, assessment and collecting different information such as:

1- The tug seaworthiness and the extend of the damage.
2- Kind of assistant required for the tug.

3- Nature and condition of cargo, stores, bunkers, in particular hazardous goods, (type, quantity, condition).
4- Queries raised by the F-Yard teams. Such as the potation for pollution, urgency of the salvage, likelihood of the potential casualty…etc.

In parallel to this, a communication channel for exchanging related data and drawings, between F-Yard the tug's companies, is online 24/7 until the end of DD.

As a result, a DD Management Plan, including management plans is generated to handle the situation in different stages.

Communication flow chart for this emergency DD as shown in Figure [6]:

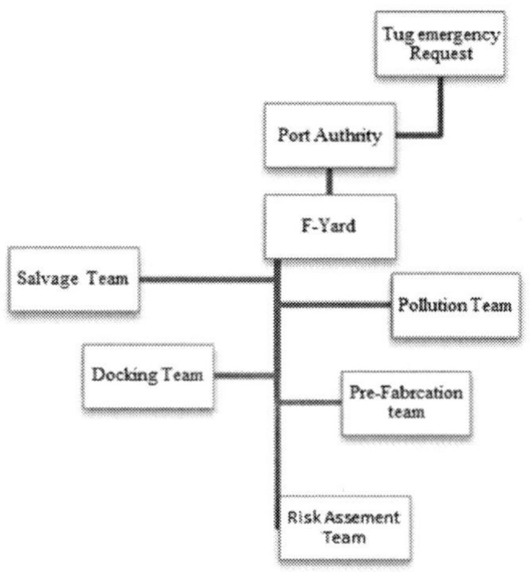

Figure [6]

While mobilization to the location, each team in the F-Yard, is focusing on its area such as pollution, prefabrication, salvage and dock preparation.

Meantime the wheelhouse is studying the required layout for the containers to optimize the production process, and the related HAZARD, HAZOP related to this DD.

7. FURTHER STUDY

This paper is a first step, drawing a framework for F-Yard concept. And there are a lot of areas that need further studies to establish F-Yard in real.

The author is preparing a full research proposal with more about the same content, to be presented for industrial use.

8. CONCLUSION

Time is an essential factor in our shipping industrial, therefore a new approach for the docking the vessels, needs to be implemented. This can be done by changing the backbone structure from a normal fixed site shipyard to a Floating Shipyard.

For this purpose, a talent management approach in management shall be used. To assure safety, productivity and quality through the whole production process.

Along with the flexibility of the F-Yard components and layout that will be tailored case by case based on the each DD requirements.

Finally, obstructions are made to be crossed by thinking without the box.

9. REFERENCES

References should be numbered in the text where they occur, with the number in square brackets [1] and a complete list given at the end of the paper.

Cross Referencing, footnotes, endnotes and 'Smart Tags' must NOT be used for figures and/or references.

1. Mossman, Alan (2009) Creating value: sufficient way to eliminate waste in lean design and lean production. *Lean Construction Journal* 2009 pp 13 – 23Bahadir
2. Dcrust Murthal, Sonepat and Haryana, 2013. An Introduction to Applicability of lean in shipbuilding. *International Journal of Latest Research in Science and Technology*, [e-journal] 2(6).Available at www.mnkjournals.com.[Accessed 2 April 2015]
3. Herrala,M.E., Pekuri,A and Aapaoja, A. 2012, How Do you understand Lean?, *20th International Group of Lean Constructions*, San Diego, USA, 10 Oct 2012 , USA, IGLC
4. Phlip c. Koenig,Hitoshi Narlta and Koichi Baba, 2002. Lean Production in the Japanese Shipbuilding Industry? *Journal of Ship Production*, 18(3), pp.167-174.
5. Richard Lee Storch ,1999, Improving flow to achieve lean manufacturing in shipbuilding, *Production Planning & Control: The Management of Operations*, 10(2), pp. 127-137
6. Thomas C. Rodzewicz,, Jonathan H. Potterton,Katherine M. Lappe, Brent C. Yezefski and David J. Singer, 2010. A new Approach to ship repair Using CONWIP and Parts Kits. *Journal of ship production and design*, 26(2), pp. 155-162.
7. Wirahadikusumah, R.D and Sulistyaningsih D., 2013. The role of owners in the supply chains of highway construction projects: An overview of Indonesian cases. *21st International Group of Lean Constructions*, Fortaleza, Brazil, 6 Sept 2013, USA, IGLC.
8. Womack, J. P., Jones,D.T., and Roos,D., 1990. The Machine that Changed the W world. New York: Rawson Associates.

9. Izunildo Fernandes Cabral. (2011) an Information Model for Lean, Agile, Resilient and Green Supply Chain Management. Master. Thesis, The Faculdade de Ciências e Tecnologia da Universidade.
10. Daniel Adjei. (2009) Application of Traditional and Agile Project Management in Consulting Firms. Master. Thesis, Umeå University.
11. Daniel I. Janson. (2016) THE DEVELOPMENT OF A GREEN SHIPYARD CONCEPT. Master. Thesis, University of Twente.
12. Martijn .L.D.Schuttevaer. (2013) Boskalis offers combined refloat and dry dock solution through SMIT and Dockwise
13. Thomas Lamb. (1986) *Engineering for ship production.* Michigan: The University of Michigan.
14. Robert Cullen. (2007) *Use of Heavy Lift Ship as a Maintenance and Repair Vessel* (MD 20817-5700). Boulevard west Bethesda: Naval Surface Warfare Center.
15. Mark V. Arena et al, (2004) *Monitoring the Progress of Shipbuilding Programs* (online). Europe: RAND Corporation. (Accessed 2005).
16. Mr. William G. Becker. (1993) *IDENTIFICATION OF NON-VALUE-ADDED TASKS IN SHIPBUILDING*
17. Ahmed Samir Ghowel. (2015) Customer's Waste at Dry-docking. In: SNAME, *world Maritime Technology Conference, 3* November. Rhode Island Convention Center and Omni Hotel.: SNAME.
18. Onno A.J. Peters. (2013) MCE Deepwater Developmen . In: The Hague, The Netherlands. *Changing the Heavy Marine Transport Market,* March. Dockwise: The Netherlands.
19. T. Terpstra, E.A. Hellinga and H.C. Leerdam. (2013) Offshore Technology Conference. In: OTC program committee, *Offshore dry-docking of FPSOs; a response to industry needs, 29–31 October.* Rio de Janeiro, Brazil: OTC.
20. D.A. Moura and R.C. Botter (2011) Sustainable Maritime Transportation and Exploitation of Sea Resources. In: Taylor & Francis Group, London, *Can a shipyard work towards lean shipbuilding or agile manufacturing.* London, September. London: ISBN.

9. AUTHORS BIOGRAPHY

Ahmed Samir Ghowel is a Maintenance Engineer in Kuwait Oil Company. Assigned to the manage the company's fleet during the dry-docking. He previously he worked in dry docks, conversion and repair projects for almost twelve years in different shipyards.

He received his BSc from Alexandria University in Naval Architecture and Marine Engineering. Thereafter, he received his MSc from Liverpool University in Construction Project Management. He was registered as Incorporated Engineer from Engineering Council-UK.

His research interests are: Lean production system and Six Sigma in Shipyard. He conducted different lecturers and presentations in different locations. He was awarded by the US Navy for his management approach in handling major repair for US Navy vessel.

AUTHORS' CONTACT DETAILS

Keith W Hutchinson
Babcock International Group - Energy and Marine Technology
Keith.W.Hutchinson@babcockinternational.com

Andrew L Scott
Maritime and Coastguard Agency (MCA)
Andrew.G.Scott@babcockinternational.com

Alex W. Vredeveldt
TNO (The Netherlands Organization)
lex.vredeveldt@tno.nl

Liza C. van Kempen
Delft University of Technology
liza.vankempen@tno.nl

Martijn G. Hoogeland
TNO (The Netherlands Organization)
martijn.hoogeland@tno.nl

Yongwon Lee
Lloyd's Register
sers@lr.org

Tobin R. McNatt
MAESTRO Marine
tobin.mcnatt@maestromarine.com

Phil Reed
Reed Maritime LLC
pr@reedmaritime.com

Rolf Catol
Western Norway University of Applied Sciences
Rolf.Cato.Imstol@hvl.no

Iain Braidwood
Ian Braidwood Consulting Ltd
iain.braidwood@navalarchitecture.ca

Robert G Allan
Robert Allan Ltd
rgallan@ral.ca

Peter Litchfield
CSD Sealing Systems
peterl@csdsealingsystems.co.uk

Stefano Persico
Lloyd's Register
Wijendra.Peiris@lr.org

Kevin F. Stone
Stone Naval Consulting
kevin@stonenaval.com

Nick Danese
NDAR
ndar@ndar.com

Rodrigo Perez
SENER, Ingeniería y Sistemas
rodrigo.fernandez@sener.es

Baiqian Jiang

London Offshore Consultants

b.jiang@loc-group.com

Paris Mangriotis

London Offshore Consultants

p.mangriotis@loc-group.com

Alexander Gorter

SMIT Salvage B.V

a.gorter@smit.com

Simon Pollard

London Offshore Consultants

s.pollard@loc-group.com

David Prentice

Lloyd's Register

David.Prentice@lr.org

Ahmed Samir

Kuwait Oil Company

aghowel@gmail.com